CAMBRIDGE TRACTS IN MATHEMATICS
General Editors
H. HALBERSTAM, C.T.C. WALL

88 *Multiple forcing*

T. JECH

Multiple forcing

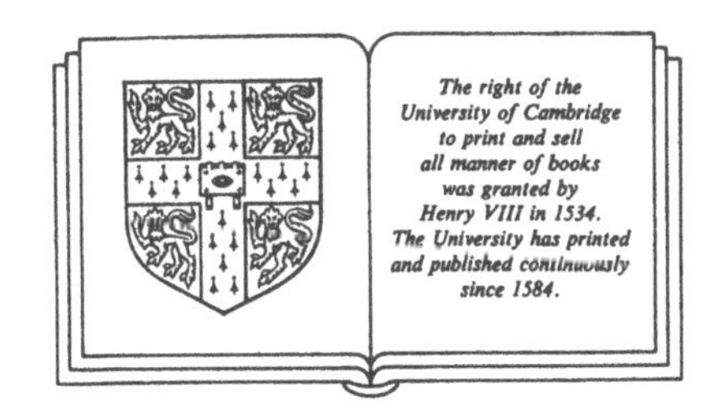
The right of the
University of Cambridge
to print and sell
all manner of books
was granted by
Henry VIII in 1534.
The University has printed
and published continuously
since 1584.

CAMBRIDGE UNIVERSITY PRESS

Cambridge

London New York New Rochelle

Melbourne Sydney

CAMBRIDGE UNIVERSITY PRESS
Cambridge, New York, Melbourne, Madrid, Cape Town, Singapore,
São Paulo, Delhi, Dubai, Tokyo, Mexico City

Cambridge University Press
The Edinburgh Building, Cambridge CB2 8RU, UK

Published in the United States of America by Cambridge University Press, New York

www.cambridge.org
Information on this title: www.cambridge.org/9780521063845

First published 1986
First paperback edition 2010

A catalogue record for this publication is available from the British Library

Jech, T.
Multiple forcing.-(Cambridge tracts in
mathematics; 88)
1. Forcing (Model theory)
I. Title
511.3 QA9.7

Library of Congress cataloguing in publication data

Jech, Thomas J.
Multiple forcing.
(Cambridge tracts in mathematics; 88)
Bibliography:
Includes indexes.
1. Forcing (Model theory) I. Title. II. Series.
QA9.7.J43 1986 511'.8 86-9601
ISBN 0 521 26659 9

ISBN 978-0-521-26659-8 Hardback
ISBN 978-0-521-06384-5 Paperback

Contents

T. JECH

Multiple forcing

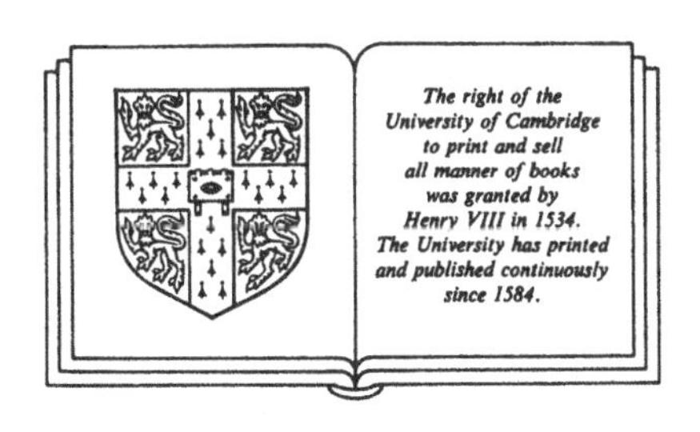
The right of the
University of Cambridge
to print and sell
all manner of books
was granted by
Henry VIII in 1534.
The University has printed
and published continuously
since 1584.

CAMBRIDGE UNIVERSITY PRESS
Cambridge
London New York New Rochelle
Melbourne Sydney

CAMBRIDGE UNIVERSITY PRESS
Cambridge, New York, Melbourne, Madrid, Cape Town, Singapore,
São Paulo, Delhi, Dubai, Tokyo, Mexico City

Cambridge University Press
The Edinburgh Building, Cambridge CB2 8RU, UK

Published in the United States of America by Cambridge University Press, New York

www.cambridge.org
Information on this title: www.cambridge.org/9780521063845

First published 1986
First paperback edition 2010

A catalogue record for this publication is available from the British Library

Jech, T.
Multiple forcing.-(Cambridge tracts in
mathematics; 88)
1. Forcing (Model theory)
I. Title
511.3 QA9.7

Library of Congress cataloguing in publication data

Jech, Thomas J.
Multiple forcing.
(Cambridge tracts in mathematics; 88)
Bibliography:
Includes indexes.
1. Forcing (Model theory) I. Title. II. Series.
QA9.7.J43 1986 511'.8 86-9601
ISBN 0 521 26659 9

ISBN 978-0-521-26659-8 Hardback
ISBN 978-0-521-06384-5 Paperback

Preface

The book is intended for a serious student of forcing. Ideally, the reader should already have some familiarity with independence proofs, as the elements of the method are only briefly reviewed in the first chapters of the book.

The main theme of the book is *multiple forcing*, a name that describes the common feature of many applications of forcing in consistency proofs. When constructing a particular generic model, one often adjoins to the existing universe a number of generic objects. This is usually done by *product forcing*, or by *iterated forcing*. An example of product forcing is Cohen's proof of independence of the continuum hypothesis, where the generic model is obtained by adjoining a large number of generic reals. An example of iterated forcing is Solovay and Tennenbaum's proof of independence of Suslin's problem, where Suslin trees are eliminated one by one, and the procedure is iterated until none remain.

The use of the method of forcing has led to many important discoveries in the two decades since Cohen's proof. The techniques have become more elaborate, often involving iteration. This led naturally to the systematic study of iterations. One result of this study was the introduction of *proper forcing* by Shelah (his book *Proper Forcing*, Springer-Verlag Lecture Notes in Mathematics 940, 1982, deals extensively with that subject).

Another result was the emergence of several *internal forcing axioms*, most notably the Proper Forcing Axiom and Martin's Maximum. These axioms, including the by now classical Martin's Axiom, enable practitioners of set theory (such as general topologists) to obtain independence results without the actual construction of a generic model.

This book attempts to give a unified treatment of various methods used over the last 20 years, and to present important applications of such methods. As the titles indicate, the first part deals with applications using product forcing and similar methods; the second part studies iteration of forcing in general; and the third part concentrates on proper forcing and related matters. A large portion of Part III is devoted to the proof of preservation of properness under countable support iteration. The proof given here is essentially due to Charles Gray (Shelah's proof can be found in the book cited above).

The book is based on a graduate course I gave in 1982–3. Eventually, I added the last two chapters dealing with more recent results. I presented a shorter version in a series of lectures in Nanjing and Beijing in May–June 1985, and am currently (during the fall of 1985) using the manuscript for a one semester course for second and third year graduate students in set theory. Needless to say that my notes have contained a number of errors and I am very grateful to all those who pointed out some of them. My hope is that not many remain in the printed version.

November, 1985

PART I

Product forcing

1 *Forcing and Boolean-valued models*

We review the basic facts of the method of forcing. The purpose is to extend the set theoretic universe V, the *ground model*, by adjoining a generic set G. We use a partially ordered set $(P, <)$, a *notion of forcing.* The elements of P are *forcing conditions*, or just conditions. A condition p is *stronger than* q if $p < q$.

A set of conditions D is *dense* (in P) if

$$\forall p \in P \quad \exists d \in D \quad d \leqslant p$$

1.1 Definition

A set $G \subset P$ is *generic* (over V) if

(i) G is a filter, i.e.
if $p \in G$ and $p \leqslant q$ then $q \in G$, and
if $p \in G$ and $q \in G$ then $\exists r \in G$ such that $r \leqslant p$ and $r \leqslant q$;
(ii) for every dense set $D \subset P$ in V, $G \cap D$ is nonempty.

Except in trivial cases, a generic set is not in the ground model. Adjoining a generic set to the ground model, we obtain a *generic extension* $V[G]$. The generic extension is a model of set theory, in fact the smallest model of set theory that extends V and contains G.

Properties of generic extensions can be described inside the ground model by means of the forcing relation. The sets in $V[G]$ have *names* in V; we shall follow the convention by using the symbol

$$\mathring{a}$$

to denote a name for the set $a \in V[G]$. The *forcing relation* is a relation defined in the ground model:

$$p \Vdash \varphi(\mathring{a}_1, \ldots, \mathring{a}_n)$$

(p *forces* φ). Here p is a condition, φ is a formula of the language of set theory, and $\mathring{a}_1, \ldots, \mathring{a}_n$ are names. The fundamental property of $\Vdash$ is the following:

1.2 The Forcing Theorem

If G is generic, then

$$V[G] \models \varphi(a_1, \ldots, a_n) \quad \text{if and only if}$$
$$\exists p \in G \quad \text{such that} \quad p \Vdash \varphi(\mathring{a}_1, \ldots, \mathring{a}_n)$$

In arguments involving forcing it is often useful to use a different version of the forcing theorem:

$$p \Vdash \varphi \quad \text{iff for every generic } G \ni p, V[G] \models \varphi$$

This version is proved from 1.2 using the assumption that for every $p \in P$ there is a generic G such that $p \in G$.

We mention two equivalent versions of genericity: A set D is *open dense* if it is dense and if

$$d \in D \quad \text{and} \quad c \leqslant d \quad \text{implies} \quad c \in D$$

A set D is *predense* if

$$\forall p \in P \quad \exists d \in D \quad \text{and} \quad \exists q \in P \quad \text{such that} \quad q \leqslant d \quad \text{and} \quad q \leqslant p$$

1.3 Proposition

A filter G is generic iff G meets every open dense set $D \in V$, iff G meets every predense set $D \in V$.

Properties of the generic extension $V[G]$ are determined not by the partial ordering $(P, <)$ itself but rather by a complete Boolean algebra $B = B(P)$ associated with P. Also, the forcing relation and the names for sets in $V[G]$ are best defined by means of the *Boolean-valued model* V^B.

First let B be a complete Boolean algebra; we use $+, \cdot, -, \sum$ and $\prod$ to denote the Boolean-algebraic operations, and $\leqslant$ for the associated partial ordering of B. Consider the partially ordered set $(B - \{0\}, <)$ as a notion of forcing. A set G is generic exactly when

(i) G is an ultrafilter on B, and either
(ii) if $A \subseteq G$ and $A \in V$ then $\prod A \in G$ or, equivalently,
(iii) if $A \in V$ and $\sum A = 1$ then $G \cap A$ is nonempty.

We say that D is *dense* in B if $D \subseteq B - \{0\}$ and is dense in $(B - \{0\}, <)$. A set $D \subseteq B - \{0\}$ is *predense* if $\sum D = 1$.

If P is dense in B and $G \subset B$ is generic, then $G \cap P$ is generic in $(P, <)$; conversely, if $H \subset P$ is generic for P then $G = \{a : \exists p \in G\, p \leqslant a\}$ is a generic ultrafilter on B. Thus, the forcing notions $B - \{0\}$ and P yield the same generic extension.

If a partially ordered set P can be embedded in a complete Boolean algebra B then generic extensions by P are the same as the generic extensions by B. Not every partial ordering can be dense in a Boolean algebra, and an additional argument is needed.

Two forcing conditions are *compatible*,

$$p|q$$

if there exists an r such that $r \leqslant p$ and $r \leqslant q$. If p and q are not compatible then they are *incompatible*:

$$p \perp q$$

In a Boolean algebra, a and b are incompatible if and only if

$$a \cdot b = 0$$

A set $A \subset P$ of mutually incompatible conditions is called an *antichain*. A *maximal antichain* is also called a *partition*.

1.4 Definition

A partially ordered set is *separative* if

$$q \nleqslant p \text{ implies that } \exists r \leqslant q, r \perp p$$

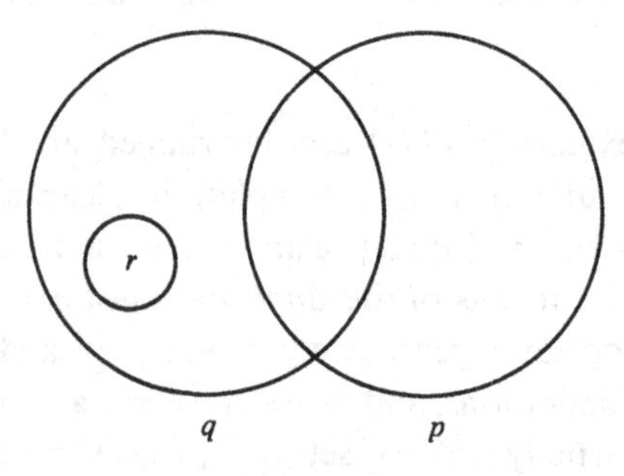

1.5 Theorem

$(P, <)$ is separative if and only if it can be embedded densely in a (unique) complete Boolean algebra.

We denote the complete Boolean algebra obtained in Theorem 1.5 by $B(P)$. If P is a Boolean algebra to begin with, $B(P)$ is its completion.

If P is an arbitrary partially ordered set, consider the equivalence relation

$$p \sim q \quad \text{if} \quad \forall r(r|p \leftrightarrow r|q)$$

The quotient of P by $\sim$, with the induced partial ordering, is a separative partial ordering, the *separative quotient of* P. It has the property that two conditions are compatible in P if and only if they are compatible in $P/\!\sim$. It follows that if $G \subset P$ is generic, then $G/\!\sim$ is generic in $P/\!\sim$, and if H is generic in $P/\!\sim$, then $G = \{p : p/\!\sim\, \in H\}$ is generic in P. Thus, $P/\!\sim$ yields the same generic models as P.

Consequently, we let $B(P) = B(P/\sim)$, and this complete Boolean algebra determines the generic extensions by P.

With a given complete Boolean algebra B we associate the *Boolean-valued model* V^B. On V^B we have two *Boolean-valued relations*

$$\|x \in y\| \quad \text{and} \quad \|x = y\|$$

that is functions from $V^B \times V^B$ into B. Using these, we define *Boolean values* of formulas

$$\|\varphi(x_1, \ldots, x_n)\|$$

where $x_1, \ldots, x_n \in V^B$, as follows:

$$\begin{aligned} \|\varphi \wedge \psi\| &= \|\varphi\| \cdot \|\psi\| \\ \|\varphi \vee \psi\| &= \|\varphi\| + \|\psi\| \\ \|\neg\varphi\| &= -\|\varphi\| \\ \|\exists x\varphi\| &= \sum_{x \in V^B} \|\varphi(x)\| \\ \|\forall x\varphi\| &= \prod_{x \in V^B} \|\varphi(x)\| \end{aligned} \tag{1.6}$$

Given a generic ultrafilter G on B, we can define the model $V[G]$ as the quotient of V^B by the ultrafilter G, i.e.

$$\begin{aligned} x/G = y/G \quad &\text{iff} \quad \|x = y\| \in G \\ x/G \in y/G \quad &\text{iff} \quad \|x \in y\| \in G \end{aligned}$$

Thus, the elements of V^B are *names* for sets in the generic extension $V[G]$. Satisfaction in $V[G]$ is described in terms of Boolean values: for any formula φ and any $\dot{x}_1, \ldots, \dot{x}_n \in V^B$,

$$V[G] \models \varphi(\dot{x}_1/G, \ldots, \dot{x}_n/G) \quad \text{iff} \quad \|\varphi(\dot{x}_1, \ldots, \dot{x}_n)\| \in G \tag{1.7}$$

The sets in the ground model have *canonical names* in V^B; we shall follow the practice of identifying the sets $x \in V$ with their canonical names. V^B also has a canonical name $\dot{G}$ for a generic ultrafilter, namely

$$\|a \in \dot{G}\| = a \qquad (a \in B)$$

Now let $(P, <)$ be again an arbitrary notion of forcing. Using the complete Boolean algebra $B = B(P)$ and the corresponding Boolean-valued model V^B and the Boolean values $\|\varphi\|$, we define the forcing relation

$$p \Vdash \varphi$$

as follows:

$$p \Vdash \varphi(\dot{x}_1,\ldots,\dot{x}_n) \quad \text{iff} \quad p \leqslant \| \varphi(\dot{x}_1,\ldots,\dot{x}_n) \|$$

where we identify p with the appropriate element of B under the embedding of (the separative quotient of) P into $B(P)$. We also define $V^P = V^{B(P)}$. The basic properties of forcing follow from (1.6):

$$\begin{array}{ll}
\multicolumn{2}{l}{\text{if } p \Vdash \varphi \text{ and } q \leqslant p \text{ then } q \Vdash \varphi} \\
\multicolumn{2}{l}{\text{there is no } p \text{ such that both } p \Vdash \varphi \text{ and } p \Vdash \neg\varphi} \\
p \Vdash \varphi \wedge \psi & \text{iff } p \Vdash \varphi \text{ and } p \Vdash \psi \\
p \Vdash \forall x \varphi & \text{iff } \forall \dot{x} \ p \Vdash \varphi(\dot{x}) \\
p \Vdash \neg\varphi & \text{iff no stronger } q \text{ forces } \varphi \\
p \Vdash \varphi \vee \psi & \text{iff } \forall q \leqslant p \ \exists r \leqslant q \ (r \Vdash \varphi \text{ or } r \Vdash \psi) \\
p \Vdash \exists x \varphi & \text{iff } \forall q \leqslant p \ \exists r \leqslant q \ \exists \dot{x} \ r \Vdash \varphi(\dot{x})
\end{array} \tag{1.8}$$

The forcing theorem 1.2 follows from (1.7). Finally, we mention this basic fact:

1.9 Proposition

For any φ and any condition p, there is a stronger condition q that *decides* φ, i.e. either $q \Vdash \varphi$ or $q \Vdash \neg\varphi$.

References

The method of forcing was invented by Cohen, see

Cohen, P. (1966). *Set Theory and the Continuum Hypothesis*, Benjamin, NY.

The detailed proofs and constructions described above can be found in:

Jech, T. (1978). *Set Theory*, Academic Press, NY.

For a different approach (not using Boolean values), see

Shoenfield, J. (1971). In *Axiomatic Set Theory* (Scott, D., ed.), pp. 357–82, American Math. Society, Providence, RI.

2 *Properties of the generic extension*

The formulas (1.7) and (1.2) describe truth in the generic extension $V[G]$ in terms of Boolean values. In practice it means that properties of $V[G]$ can be described in the ground model V in terms of properties of the Boolean algebra B (or of the forcing notion P).

First we illustrate this principle on a simple example:

2.1 Proposition

B is atomless if and only if the generic ultrafilter is not in the ground model.

(Precisely: for every ultrafilter G on B that is generic over V, $G \notin V$.)

Proof If a is an atom of B, then the ultrafilter $\{b \in B : b \geqslant a\}$ is generic and is in V. Conversely, let F be an ultrafilter on B, generic over V. If $F \in V$ and if B is atomless, then the set $B - F$ is dense, and so it meets F; a contradiction. □

As the next example, we consider the relationship between automorphisms of B and definability. B is *weakly homogeneous* if for any nonzero u and v there is an automorphism π of B such that $\pi(u) \cdot v \neq 0$.

An automorphism of B extends to an automorphism of the Boolean-valued model V^B via

$$\pi(\| x \in y \|) = \| \pi(x) \in \pi(y) \| \quad \text{and} \quad \pi(\| x = y \|) = \| \pi(x) = \pi(y) \|$$

All canonical names for sets in V are preserved under all automorphisms. Thus, if φ is a formula with parameters from V and π is any automorphism of B, we have

$$\pi(\| \varphi \|) = \| \varphi \|.$$

Now if B is weakly homogeneous, the only elements of B fixed under all automorphisms are 0 and 1, and so the Boolean value of any formula with free variables in V is either 0 or 1. Thus, we have proved the following:

2.2 Proposition

Let B be weakly homogeneous. If $S \in V[G]$ is a subset of V and is definable in $V[G]$ by a formula with parameters in V, then $S \in V$. □

Now we turn our attention to submodels of the generic extension. Let B be a complete Boolean algebra and let A be a complete Boolean subalgebra of B. If G is a generic ultrafilter on B then $G \cap A$ is easily seen to be a generic ultrafilter on A. Clearly, the generic extension $V[G \cap A]$ is a submodel of $V[G]$, and in fact

$$V \subseteq V[G \cap A] \subseteq V[G]$$

We shall now show that, conversely, all submodels of $V[G]$ that extend V are obtained that way:

2.3 Lemma

If $V \subseteq M \subseteq V[G]$ and M is a model of ZFC then there exists a complete Boolean subalgebra A of B such that $M = V[G \cap A]$.

Proof Let Z be the power set of B in M. Because M is a model, there exists a set of ordinals $S \in M$ such that $Z \in V[S]$.

Now if X is any set of ordinals in $V[G]$, let A_X be (in V) the complete subalgebra of B generated by $\{\|\alpha \in \dot{X}\| : \alpha \text{ an ordinal}\}$. It is not difficult to show that $G \cap A_X \in V[X]$ and $X \in V[G \cap A_X]$; hence $V[X] = V[G \cap A_X]$.

We claim that $M = V[G \cap A_S]$. To prove that, it suffices to show that every set $X \in M$ of ordinals is in $V[S]$. But if $X \in M$ is a set of ordinals then $G \cap A_X \in Z$ and so $X \in V[Z] \subseteq V[S]$. It follows that $M = V[S] = V[G \cap A_S]$. □

When dealing with notions of forcing, it is not always easy to recognize when $B(P)$ is a complete Boolean subalgebra of $B(Q)$. Instead, we state below some sufficient conditions for V^P to be a submodel of V^Q. The precise meaning of $V^P \subseteq V^Q$ is that whenever G is a generic filter on Q then there is $H \in V[G]$ which is a generic filter on P.

First we look at the case when $(P, <) \subseteq (Q, <)$; that is $P \subseteq Q$ and the partial orderings agree.

2.4 Lemma

Let $P \subset Q$ be such that

(i) for all $p_1, p_2 \in P$, if p_1 and p_2 are compatible in Q then they are compatible in P, and

(ii) every maximal antichain in P is maximal in Q.

Then $V^P \subseteq V^Q$.

Proof If G is a generic filter on Q then $G \cap P$ is a generic filter on P. □
(Note that (ii) implies that P is predense in Q.)

In the next lemma, its condition (ii) implies (ii) of 2.4:

2.5 Lemma
Let $P \subset Q$ be such that

(i) for all $p_1, p_2 \in P$ if $p_1 |_Q p_2$ then $p_1 |_P p_2$, and
(ii) $\forall q \in Q \exists p \in P$ such that all $p' \leqslant p$ in P are compatible with q.

Then $V^P \subseteq V^Q$. □

We can weaken the assumption $P \subset Q$:

2.6 Lemma
Let $i{:}P \to Q$ be such that

(i) if $p_1 \leqslant p_2$ then $i(p_1) \leqslant i(p_2)$,
(ii) for all $p_1, p_2 \in P$, if $p_1 \perp p_2$ then $i(p_1) \perp i(p_2)$, and
(iii) $\forall q \in Q \exists p \in P$ such that $\forall p' \leqslant p$, $i(p') | q$.

Then $V^P \subseteq V^Q$.

Proof If G is a generic filter on Q then $i^{-1}(G) \cap P$ is a generic filter on P. □

We state one more sufficient condition:

2.7 Lemma
Let $h{:}Q \to P$ be such that

(i) if $q_1 \leqslant q_2$ then $h(q_1) \leqslant h(q_2)$, and
(ii) $\forall q \in Q \forall p' \leqslant h(q) \exists q'$ such that $q' | q$ and $h(q') \leqslant p'$.

Then $V^P \subseteq V^Q$.

Proof Note that if $D \subseteq P$ is open dense then $h^{-1}(D)$ is predense in Q. It follows that if G is generic on Q then $\{p \in P{:}p \geqslant h(q)$ for some $q \in G\}$ is generic on P. □

The relationship between properties of the forcing notion and properties of the generic extension is perhaps best illustrated by *distributivity*.

2.8 Definition

Let κ be an infinite cardinal. $(P, <)$ is *κ-distributive* if the intersection of any κ open dense sets is dense.

An equivalent formulation is in terms partitions. A partition W_1 of P is a *refinement* of a partition W_2 if $\forall p_1 \in W_1 \exists p_2 \in W_2$ such that $p_1 \leqslant p_2$. We leave the proof of the equivalence to the reader:

2.9 Proposition

$(P, <)$ is κ-distributive if and only if any κ partitions of P have a common refinement. □

A complete Boolean algebra is κ-distributive iff it satisfies the *distributive law*

$$\prod_{\alpha<\kappa}\sum_{i\in I} u_{\alpha i} = \sum_{f:\kappa\to I}\prod_{i\in I} u_{\alpha, f(\alpha)}$$

2.10 Theorem

$(P, <)$ is κ-distributive if and only if every function $f:\kappa \to V$ in the generic extension is in the ground model V.

Proof (*a*) Let $(P, <)$ be κ-distributive, let G be generic, and let $f \in V[G]$ be a function from κ into V. There is $p_0 \in G$, a name $\dot{f}$ and $Y \in V$ such that

$$p_0 \Vdash \dot{f} \text{ is a function from } \kappa \text{ into } Y$$

For each $\alpha < \kappa$ we let

$$D_\alpha = \{p\colon \text{either } p \perp p_0, \text{ or } \; p \leqslant p_0 \; \text{ and } \; \exists y \in Y \text{ such that } p \Vdash \dot{f}(\alpha) = y\}$$

Each D_α is open dense, and so $D = \bigcap_{\alpha<\kappa} D_\alpha$ is dense. By genericity, there is $p \in G$ such that $p \in D$. For each $\alpha < \kappa$ let $g(\alpha)$ be the y such that $p \Vdash \dot{f}(\alpha) = y$. The function g is in V, and $p \Vdash \dot{f} = g$, therefore $f = g$.

(*b*) Conversely, assume that $V[G]$ has the property stated in the theorem, and let W_α, $\alpha < \kappa$, be partitions. Consider the name $\dot{f}$ for a function on κ such that

$$\| \dot{f}(\alpha) = u \| = u$$

for all $u \in W_\alpha$. If p is any condition, then there exists $g \in V$ and some $q \leqslant p$ such that $q \Vdash \dot{f} = g$. It follows that for every α there exists $u \in W_\alpha$ such that $q \leqslant u$. Consequently, there is a partition that refines all the W_α. □

A sufficient condition for P to be κ-distributive is when P is *κ-closed*: For every $\lambda \leqslant \kappa$, every descending sequence

$$p_0 \geqslant p_1 \geqslant \cdots \geqslant p_\alpha \geqslant \cdots \qquad (\alpha < \lambda)$$

has a lower bound.

2.11 Lemma

If $(P, <)$ is κ-closed then it is κ-distributive.

Proof Exercise. □

Theorem 2.10 can be made more specific: A *complete* Boolean algebra is (κ, λ)-distributive, if any κ partitions of size $\leqslant \lambda$ have a common refinement.

2.12 Theorem

B is (κ, λ)-distributive if and only if every function $f:\kappa \to \lambda$ in $V[G]$ is in V.

Proof Similar. □

Another important property used in forcing is the *chain condition.*

2.13 Definition

$(P, <)$ satisfies the *κ-chain condition* if every partition has size less than κ.

P satisfies the κ-chain condition if and only if $B(P)$ does. The ω_1-chain condition is traditionally called the *countable chain condition* (c.c.c.). We also say that P is *κ-saturated* if it has the κ-c.c., and denote

$$\text{sat}(P)$$

the least κ such that P is κ-saturated. (It is always a regular uncountable cardinal.)

Note that if A is subalgebra of B then $\text{sat}(A) \leqslant \text{sat}(B)$.

2.14 Theorem

Let κ be a regular cardinal, and let $(P, <)$ satisfy the κ-chain condition. Then κ is a regular cardinal in $V[G]$.

Proof It suffices to prove that for every $\lambda < \kappa$, every function $f:\lambda \to \kappa$ is bounded below κ. Let $\dot{f}$ be a name for such an f; for each $\alpha < \lambda$ let $A_\alpha = \{\beta < \kappa : \| \dot{f}(\alpha) = \beta \| \neq 0\}$. As the values $\| \dot{f}(\alpha) = \beta \|$ constitute a partition, we have $|A_\alpha| < \kappa$. Since κ is regular and $\lambda < \kappa$, all the A_α are bounded below κ by the same bound. It follows that f is bounded below κ. □

Theorems 2.10 and 2.14 are useful in practice. All cardinals in $V[G]$ are of course cardinals in V, but a cardinal in V need not remain a cardinal in $V[G]$. Distributivity and chain conditions give a sufficient condition for preservation of cardinals:

2.15 Corollary

Let $\lambda \leqslant \kappa$ be regular cardinals. If P is γ-distributive for every $\gamma < \lambda$ and has the κ-chain condition then all cardinals $\leqslant \lambda$ and all cardinals $\geqslant \kappa$ (in V) remain cardinals in $V[G]$. If P has the countable chain condition then all cardinals are preserved.

The following lemma gives an estimate of the number of reals in a generic extension:

2.16 Lemma

If P has the κ-chain condition, then $(2^{\aleph_0})^{V[G]} \leqslant (|P|^{<\kappa})^V$.

Proof Each real $a \in V[G]$ with a name $\mathring{a}$ is determined by a function $n \to \| n \in \mathring{a} \|$ and so $(2^{\aleph_0})^{V[G]} \leqslant |B|^{\aleph_0}$, where $B = B(P)$. But $|B| \leqslant |P|^{<\kappa}$ if sat $P \leqslant \kappa$. □

In particular, if P has the c.c.c., then $(2^{\aleph_0})^{V[G]} \leqslant |P|^{\aleph_0}$.

References

The properties discussed are standard in the theory of forcing; details can be found in textbooks, for example

Jech, T. (1978). *Set Theory*, Academic Press, NY.
Kunen, K. (1980). *Set Theory*, North-Holland, Amsterdam.

3 Examples of generic reals

We describe several ways of adjoining a new *real*, that is a subset of ω (or a function from ω into $\{0, 1\}$, or a function from ω into ω).

3.1 Cohen reals

The notion of forcing $(P, <)$ consists of finite sequences of 0's and 1's. A condition $p \in P$ is stronger than $q \in P$ if $p \supseteq q$ (*p extends q*).

If G is a generic set of conditions, let

$$f = \bigcup \{p : p \in G\}$$

and

$$a = \{n : f(n) = 1\}$$

The function f, or the set $a \subset \omega$, is called a *Cohen real*. An easy forcing argument shows that a Cohen real is not in the ground model. Moreover, $M[G] = M[f] = M[a]$, because G can be recovered from f, namely $G = \{p \in P : p \subset f\}$.

The set P is countable, and so it satisfies the c.c.c., and so $V[G]$ preserves all cardinals. The algebra $B(P)$ is the unique atomless complete Boolean algebra that has a countable dense subset (*separable*).

The same model $V[G]$ is obtained by forcing with the forcing notion $(Q, <)$ whose elements are finite sequences of natural numbers (finite partial function from ω into ω). If G is a generic subset of Q then the function

$$f = \bigcup \{p : p \in G\}$$

(a function from ω into ω) is also called a Cohen real.

If A is subalgebra of $B = B(P)$ then since A is separable, the model V^A is isomorphic to V^B.

3.2 Random reals

The forcing conditions are Borel sets of positive Lebesgue measure in R (or in 2^ω). A condition P is stronger than q if $p \subseteq q$. The corresponding Boolean algebra B is the algebra of Borel sets modulo the sets of measure zero. B is a measure algebra, and is the unique countably generated measure algebra.

The generic extension $V[G]$ is determined by a single real, a *random real*. Let $a \in R^{V[G]}$ be the unique member of each rational interval $(r_1, r_2)^{V[G]}$ such that $(r_1, r_2)^V \in G$. Conversely, G can be recovered from a, and so $V[G] = V[a]$.

The measure algebra B satisfies the c.c.c., and so $V[G]$ preserves all cardinals.

Every subalgebra A of B is again a measure algebra, and so V^A is isomorphic to V^B.

The following lemma exhibits a difference between the Cohen and random reals. If f and g are functions from ω into ω, then f *dominates* g if $f(n) > g(n)$ for all n.

3.3 Lemma

(*a*) In the random extension, every $f:\omega \to \omega$ is dominated by some $g \in V$.

(*b*) In the Cohen extension it is not so.

Proof (*a*) Let p be a condition, and $p \Vdash \dot{f}:\omega \to \omega$; we find a $q < p$ and some g such that $q \Vdash g$ dominates $\dot{f}$. Let μ be the measure on B.

First we find a number $k = g(0)$ such that

$$\mu(p - \| \dot{f}(0) < k \|) < \tfrac{1}{4} \cdot \mu(p)$$

This is possible because $\{ \| \dot{f}(0) = j \| : j \in \omega\}$ is a partition of p. Similarly we find for each n a number $g(n)$ such that

$$\mu(p - \| \dot{f}(n) < g(n) \|) < \frac{1}{2^n} \cdot \tfrac{1}{4} \mu(p)$$

It follows that the Borel set

$$q = p \cap \bigcap_{n=0}^{\infty} \| \dot{f}(n) < g(n) \|$$

has positive measure (at least $\frac{1}{2}\mu(p)$), and therefore is a condition. It forces $\forall n \dot{f}(n) < g(n)$.

(*b*) Let $f:\omega \to \omega$ be a Cohen real; for any $p \in Q = (\omega^{<\omega}, \supseteq)$

$$p \Vdash p \subset f$$

If $g:\omega \to \omega$ is in V then there exists $q \supset p$ and $n \in \operatorname{dom}(q)$ such that $q(n) > g(n)$. It follows that

$$q \Vdash g \text{ does not dominate } \dot{f} \quad \square$$

3.4 Sacks reals

The forcing uses perfect sets of reals, ordered by inclusion. An equivalent formulation uses *perfect trees.*

Definition A *tree* (a $0-1$-tree of height $\leqslant \omega$) is a subset p of $\{0,1\}^{<\omega}$ with the property

$$\text{if} \quad s \in p \quad \text{and} \quad t \subset s \quad \text{then} \quad t \in p$$

A tree p is *perfect* if $\forall s \in p$ there is $t \supseteq s$ such that both $t^\frown 0 \in p$ and $t^\frown 1 \in p$.

The set of all paths in a perfect tree is a perfect set in $\{0,1\}^\omega$.

The forcing notion $(P, <)$ consists of all perfect trees. The ordering is by inclusion: $p < q$ iff $p \subseteq q$.

If G is generic, let

$$f = \bigcup \{s : \forall p \in G \quad s \in p\}$$

$f : \omega \to \{0,1\}$ is a *Sacks real.* Again, $V[G] = V[f]$, because $G = \{p \in P : \forall n \quad f \restriction n \in p\}$.

Sacks forcing does not have the c.c.c. Since $|P| = 2^{\aleph_0}$, P has the $(2^{\aleph_0})^+$-chain condition, and so if we assume $2^{\aleph_0} = \aleph_1$ in the ground model, then sat $(P) = \aleph_2$, and so all cardinals $\geqslant \aleph_2$ are preserved.

3.5 Lemma

Let X be a set of ordinals in $V[G]$. If

$$p \Vdash \dot{X} \text{ is countable}$$

then there exists a countable set A and some $q \leqslant p$ such that

$$q \Vdash \dot{X} \subseteq A$$

It follows that $\aleph_1$ is preserved in $V[G]$. The proof of Lemma 3.5 uses the technique of *fusion.*

A node $s \in p$ is an nth *branching point* of a tree p if both $s^\frown 0$ and $s^\frown 1$ are in p, and if there are exactly n nodes $t \subseteq s$ such that $t^\frown 0 \in p$, $t^\frown 1 \in p$. A perfect tree has 2^{n-1} nth branching points for each n.

3.6 Definition

$p \leqslant_n q$ means that $p \leqslant q$, and every nth branching point of q is a branching point of p. (And $p \leqslant_0 q$ is just $p \leqslant q$.)

A *fusion sequence* is a sequence such that

$$p_0 \geqslant_0 p_1 \geqslant_1 p_2 \geqslant_2 \cdots \geqslant_{n-1} p_n \geqslant_n \cdots$$

3.7 Lemma

If $\{p_n\}_n$ is a fusion sequence then $p=\bigcap_{n=0}^{\infty}p_n$ (the *fusion* of $\{p_n\}$) is a perfect tree.

Moreover, $p\leqslant_n p_n$ for every n.

Proof Easy. □

For $s\in p$, let

$$p\upharpoonright s=\{t\in p: t\subseteq s \quad \text{or} \quad t\supseteq s\}$$

Let p be a condition, and let $n\geqslant 1$. Let s be an nth branching point of p. Let $t=s\frown 0$ or $t=s\frown 1$, and let q be such that $q\leqslant p\upharpoonright t$. Then

$$r=q\cup\{u\in p: u\nsubseteq t \quad \text{and} \quad t\nsubseteq u\} \tag{3.8}$$

is a condition, the *n-amalgamation* of q into p. And $r\leqslant_n p$.

More generally, if $t_1,\dots,t_k$ are all the distinct successors of nth branching points, and $q_i\leqslant p\upharpoonright t_i$ for $i=1,\dots,k$, then the amalgamation of $\{q_1,\dots,q_k\}$ into p is $q_1\cup\cdots\cup q_k$.

Now we are ready to prove Lemma 3.5:

Proof Let p be a condition, and $p\Vdash \dot{X}:\omega\to V$. First we find $p_0\leqslant p$ and $A_0=\{a_0\}$ such that

$$p_0\Vdash \dot{X}(0)=a_0$$

We proceed by induction. At stage n, we have already constructed p_{n-1}. Let $t_i, i=1,\dots,k=2^n$, be all the $s\frown 0$ and $s\frown 1$, where s ranges over the nth branching points of p_{n-1}. For each $i\leqslant k$, there exists $q_i\leqslant p_{n-1}\upharpoonright t_i$ and some a_n^i such that

$$q_i\Vdash \dot{X}(n)=a_n^i$$

Let $A_n=\{a_n^i: i=1,\dots,k\}$. Let p_n be the amalgamation of $\{q_i\}_{i\leqslant k}$ into p_{n-1}. We have $p_n\leqslant_n p_{n-1}$, and

$$p_n\Vdash \dot{X}(n)\in A_n$$

The sequence $p\geqslant_0 p_0\geqslant_1 p_1\geqslant_2\dots$ is a fusion sequence; let q be its fusion. Let $A=\bigcup_{n=0}^{\infty}A_n$. It follows that $q\Vdash \text{range}(\dot{X})\subseteq A$. □

Using the technique of fusion, one can prove the following theorem that we state here without proof:

3.9 Theorem (Sacks)

A Sacks real f is minimal over V; i.e. if $X \in V[f]$ is a set of ordinals then either $V[X] = V$ or $V[X] = V[f]$. □

3.10 Prikry–Silver reals

This forcing is similar to the Sacks forcing. A condition $p \in P$ is a function with values 0 and 1, defined on a co-infinite subset of ω. A condition p is stronger than q if p extends q.

A *Prikry–Silver* real is the function $f:\omega \to \{0,1\}$

$$f = \bigcup \{p : p \in G\}$$

where G is a generic filter on P.

As in the case of Sacks forcing, P has size $2^{\aleph_0}$ and does not satisfy the c.c.c P preserves $\aleph_1$; indeed it satisfies Lemma 3.5. The argument is similar to the one for Sacks forcing, and in fact a Prikry–Silver real is minimal.

A *fusion* for Prikry–Silver forcing is defined by defining $\leqslant_n$:

$p \leqslant_n q$ iff $p \supseteq q$ and the first n elements not in the domain of q are $\notin \operatorname{dom}(p)$

If $\{p_n\}$ is a fusion sequence, then $p = \bigcup_{n=0}^{\infty} p_n$ (the *fusion*) is a condition, and $p \leqslant_n p_n$ for all n.

The proof of Lemma 3.5 for P is similar to the proof for Sacks forcing, just a little more complicated. There is also a way of looking at P that makes the analogy more apparent. Let us call a perfect tree T *uniform* if for any $s \in T$ and $t \in T$ of the same length, we have

$$s^\frown 0 \in T \text{ iff } t^\frown 0 \in T, \quad s^\frown 1 \in T \text{ iff } t^\frown 1 \in T$$

The forcing P is isomorphic to the set of all uniform perfect trees. Given a condition $p \in P$, we define a uniform T as follows: when $s \in T$, $s \in \{0,1\}^n$, then $s^\frown p(n) \in T$ if $p(n)$ is defined, and both $s^\frown 0 \in T$ and $s^\frown 1 \in T$ if $p(n)$ is undefined.

The proof of Lemma 3.5 goes through for uniform trees with the only exception. When we construct p_n from p_{n-1}, we cannot amalgamate the q_i's simultaneously. In order to obtain a uniform tree, we have to do it successively (in 2^n steps), choosing q_i at the ith step.

3.11 Mathias reals

A forcing condition is a pair (s, S), where s is a finite increasing sequence of natural numbers, and S is an infinite subset of ω-$\max(s)$. The intended generic object, a *Mathias real*, is an increasing $f:\omega \to \omega$ such that

$$s \subset f \subset s \cup S$$

(here we freely confuse sequences with their range). This motivates the partial ordering

$$(t, T) \leqslant (s, S) \quad \text{if} \quad t \supseteq s, T \subseteq S, \text{ and range } (t)\text{-range } (s) \subset S$$

The finite sequence s is called the *stem* of the condition (s, S).

If G is a generic set, the corresponding Mathias real is the function

$$f = \bigcup \{s{:}(s, S) \in G \text{ for some } S\}$$

We have $V[f] = V[G]$, as

$$G = \{(s, S){:}s \subset f \subset s \cup S\}$$

Mathias forcing satisfies the $(2^{\aleph_0})^+$-chain condition, but not the c.c.c. It does preserve $\aleph_1$, however. The proof is analogous to the proof of Lemma 3.5 for the perfect set forcing, but with an added twist. For each n, let us define

$$q \leqslant_n p \quad \text{iff}$$

(i) p and q have the same stem, $p = (s, S)$, $q = (s, T)$ and $S \supset T$, and
(ii) the first n elements of S are in T.

If $p_0 \geqslant_0 p_1 \geqslant_1 p_2 \geqslant_2 \ldots$ is a fusion sequence, $p_n = (s, S_n)$, then $p = (s, \bigcap_{n=0}^{\infty} S_n)$ is a condition (the *fusion*), has the same stem, and $p \leqslant_n p_n$ for all n.

3.12 Lemma

If $p \Vdash \dot{X}{:}\omega \to V$ then there exists $q \leqslant p$ with the same stem, and a countable A such that $q \Vdash \dot{X} \subseteq A$.

Proof Let $\{u_n\}_n$ be a sequence of natural numbers such that each u appears infinitely often. We construct a fusion sequence $\{p_n\}$ with $p_0 = p$, and a sequence of sets A_n with the intention that the fusion forces $\dot{X} \subseteq \bigcup_{n=0}^{\infty} A_n$.

At stage n, we already have p_n and A_{n-1}. Let $p_n = (s, S_n)$, where s is the stem of p. Let K_n be the set consisting of the first n elements of S_n. Let $t_k, k = 1, \ldots, 2^n$, be all the possible extensions of s such that range $(t_k - s) \subseteq K_n$. We construct $p_{n+1} = (s, S_{n+1})$ in 2^n steps, by successively constructing $S_n = S^{(0)} \supseteq S^{(1)} \supseteq \cdots S^{(k)} \supseteq \cdots \supseteq S^{(2^n)} = S_{n+1}$.

At stage k, if there exists $T_k \subseteq S^{(k-1)} - K_n$ and some a_n^k such that

$$(t_k, T_k) \Vdash \dot{X}(u_n) = a_n^k \tag{3.13}$$

then we let $S^{(k)} = K_n \cup T_k$. We also put a_n^k into A_n. We call the resulting condition $(s, S^{(k)})$ the *n-amalgamation* of (t_k, T_k) into $(s, S^{(k-1)})$; note that $(s, S^{(k)}) \leqslant_n (s, S^{(k-1)})$. If no T_k with property (3.13) exists, we let $S^{(k)} = S^{(k-1)}$.

In 2^n steps we obtain a condition $p_{n+1} = (s, S_{n+1})$; we have $p_{n+1} \leqslant_n p_n$. We have also constructed a set A_n.

We let $p_\infty = (s, S)$ be the fusion of $\{p_n\}$, and $A = \bigcup_{n=0}^{\infty} A_n$. We want to show that $p_\infty \Vdash \dot{X} \subseteq A$. Thus, let $q \leqslant p_\infty$ be any condition, and let $u \in \omega$; we find $r \leqslant q$ such that $r \Vdash \dot{X}(u) \in A$.

There is a condition $(t, T) \leqslant q$ and some a such that $(t, T) \Vdash \dot{X}(u) = a$. Let n be large enough so that $u = u_n$, and that the range of $t - s$ is included among the first n elements of S. As $p_\infty \leqslant_n p_n$, the first n elements of S are exactly K_n. And t is one of the t's considered at stage n; say $t = t_k$. Now $t, T - K_n, a$ satisfy (3.13) and so we have chosen some T_k and a_n^k at that stage. But $T - K_n \subseteq S - K_n \subseteq T_k$ and so $r = (t, T - K_n)$ is stronger than both (t, T) and (t, T_k); it follows that $a = a_n^k$; hence, $a \in A$ and $r \Vdash \dot{X}(u) \in A$. □

An important property of Mathias forcing is stated (without proof) in this lemma:

3.14 Lemma

For any condition p and any sentence σ of the forcing language there is a condition $q \leqslant p$ with the same stem which decides σ. □

One consequence of Lemma 3.14 is that any infinite subset of a Mathias real is a Mathias real.

Note that Mathias reals grow faster than all ground model reals:

3.15 Lemma

If f is a Mathias real then any $g: \omega \to \omega$ in V is eventually dominated by f, i.e. $g(n) < f(n)$ for all but finitely many n.

Proof Let $g \in V$ and let $p = (s, S)$ be any condition. There is an infinite $T \subseteq S$ such that for every $n > k = |s|$, then nth element of $s \cup T$ is greater than $g(n)$. The condition (s, T) forces that the generic real dominates g because every f such that $s \subset f \subset s \cup T$ does. □

3.16 Laver reals

A set $p \subseteq \omega^{<\omega}$ is a *tree* if it is closed under initial segments. A tree p is a *Laver tree* if it has $s \in p$ (the *stem* of p) such that

(i) $\forall t \in p$ either $t \subseteq s$ or $t \supseteq s$

(ii) $\forall t \supseteq s$ the set $S(t) = \{a : t^\frown a \in p\}$ of all successors of t in p is infinite.

Laver forcing consists of all Laver trees, partially ordered by inclusion.

If G is a generic set then

$$f=\bigcup\{s: s \text{ is the stem of some } p\in G\}$$

is a function from ω into ω, a *Laver real.* Since $G=\{p: \text{stem}(p)\subset f$ and $\forall n\geqslant|s|\ f(n)\in S(f\upharpoonright n)\}$, we have $V[G]=V[f]$. Similar to Lemma 3.15, we have

3.17 Lemma

A Laver real eventually dominates every $g:\omega\to\omega$ in V.

Proof Given g and a condition p with stem s, we let $q=\{t\in p: \forall n\geqslant|s|\ t(n)>g(n)\}$. The condition q forces that $\mathring{f}(n)>g(n)$ for all $n\geqslant|s|$.

Laver forcing satisfies the $(2^{\aleph_0})^+$-chain condition, but not the c.c.c. The proof that $\aleph_1$ is preserved is similar to the proof for Mathias reals.

Consider a canonical enumeration of $\omega^{<\omega}$ in which s appears before t if $s\subset t$, and $s\frown a$ appears before $s\frown(a+1)$. If p is a Laver tree, then the part of p above the stem is isomorphic to $\omega^{<\omega}$, and so we have an enumeration

$$s_0^p=\text{stem}(p),\ s_1^p,\ldots,s_n^p,\ldots$$

of $\{t\in p: t\supseteq\text{stem}(p)\}$, for every Laver tree. Note that if $q\leqslant p$ and $s=s_n^q=s_m^p$ then $n\leqslant m$. Let

$$q\leqslant_n p \text{ if } \text{stem}(q)=\text{stem}(p), \text{ and } s_i^p=s_i^q \text{ for all } i=0,\ldots,n$$

3.18 Lemma

If $p_0\geqslant_0 p_1\geqslant_1 p_2\geqslant_2\cdots\geqslant_n\cdots$(a *fusion sequence*), then $p=\bigcap_{n=0}^{\infty}p_n$ is a Laver tree (the *fusion* of $\{p_n\}_n$), with the same stem, and $p\leqslant_n p_n$ for all n.

Proof Let s_0 be the stem of p_0. The set $S^p(s_0)=\bigcap_n S^{p_n}(s_0)$ is infinite. For every $a\in S^p(s_0)$, $S^p(s_0\frown a)=\bigcap_n S^{p_n}(s_0\frown a)$ is infinite, and so on. □

Let p be a Laver tree and let $n\geqslant 0$. Let t be a $\subseteq$-maximal node among $\{s_0^p,\ldots,s_n^p\}$, and let $q\leqslant p$ be a Laver tree with stem t. Then

$$r=q\cup\{u\in p: u\nsubseteq t \text{ and } t\nsubseteq u\} \tag{3.19}$$

is a condition, the *n-amalgamation* of q into p. And $r\leqslant_n p$. Moreover (as with Sacks forcing), when $t_1,\ldots,t_k$ are all the maximal nodes among $\{s_0^p,\ldots,s_n^p\}$, and $q_1,\ldots,q_k$ are trees $\leqslant p$ with stems $t_1,\ldots,t_k$, the amalgamation of $\{q_1,\ldots,q_k\}$ into p is the obvious generalization of (3.19).

3.20 Lemma

If $p \Vdash \dot{X}:\omega \to V$, then there is $q \leqslant p$ with the same stem, and a countable A such that $q \Vdash \dot{X} \subseteq A$.

Proof Let $\{u_n\}_n$ be a sequence of natural numbers such that each u appears infinitely often. We construct a fusion sequence $\{p_n\}$ with $p_0 = p$, and sets A_n so that the fusion forces $\dot{X} \subseteq \bigcup_n A_n$.

At stage n, we already have p_n. Let $t_1, \ldots, t_k$ be all the maximal nodes among $\{s_0^{p_n}, \ldots, s_n^{p_n}\}$. For each $i = 1, \ldots, k$, if there exists $q_i \leqslant p_k$ with stem t_i and a_n^i so that

$$q_i \Vdash \dot{X}(u_n) = a_n^i \tag{3.21}$$

we choose such q_i and a_n^i. We let A_n be the collection of all the a_n^i, and let p_{n+1} be the amalgamation of $\{q_1, \ldots, q_k\}$ into p_n. We have $p_{n+1} \leqslant_n p_n$.

Let p_∞ be the fusion of $\{p_n\}_n$; p_∞ has the same stem as p_0. Let $A = \bigcup_{n=0}^{\infty} A_n$. To prove that $p_\infty \Vdash \dot{X} \subseteq A$, let $q \leqslant p_\infty$, and let $u \in \omega$. There is a condition $\bar{q} \leqslant q$ and some a such that $\bar{q} \Vdash \dot{X}(u) = a$. Let n be large enough so that $u = u_n$ and that the stem of $\bar{q}$ is among $K = \{s_0^{p_n}, \ldots, s_n^{p_n}\}$. There is $t \in \bar{q}$ that is a maximal node in K; it is one of the nodes we considered at stage n, say $t = t_i$. Let $r = \bar{q} \restriction t = \{v \in \bar{q} : v \subseteq t \text{ or } v \supseteq t\}$. As r, a satisfy (3.21), we have chosen q_i and a_n^i. Because $r \leqslant q_i$, it must be the case that $a = a_n^i$, and so $r \Vdash \dot{X}(u) \in A$.

Hence, $p_\infty \Vdash \dot{X} \subseteq A$. □

Mathias forcing and Laver forcing have similar properties. The following formulation of Mathias forcing makes the similarity more explicit.

Let us call a Laver tree p with stem s *uniform* if s is increasing and if there exists $S \subseteq \omega$ such that for every $t \supseteq s$ in p, $S(t) = \{a \in S : a > \max(t)\}$. It is clear that p corresponds to the Mathias forcing condition (s, S), and so Mathias forcing is isomorphic to the set of all uniform Laver trees.

3.22 Grigorieff reals

We mention this forcing without proofs. The fusion argument for Grigorieff forcing uses a fusion tree rather than a fusion sequence and is somewhat more complicated.

Let D be a nonprincipal ultrafilter on ω. D is called a *p-point* if for any countable collection $\{A_n : n \in \omega\} \subset D$ there exists some $A \in D$ such that for every n, $A - A_n$ is a finite set.

Let D be a p-point. A forcing condition is a function p into $\{0, 1\}$ with the property that $\operatorname{dom}(p) \notin D$. A condition q is stronger than p if $q \supseteq p$.

A Grigorieff real is a function

$$f = \bigcup\{p : p \in G\}$$

We also have $G = \{p : p \subset f\}$.

Grigorieff forcing has size $2^{\aleph_0}$ but does not satisfy the c.c.c. However,

3.23 Lemma

Grigorieff forcing preserves $\aleph_1$. □

It should be noted that the assumption that D is a p-point is necessary: if D is not a p-point then Grigorieff forcing collapses $\aleph_1$. Also, the existence of p-points follows from the continuum hypothesis (CH).

References

Sacks reals:

Sacks, G. (1971). In *Axiomatic Set Theory* (Scott, D., ed.), pp. 331–55. American Math. Society Providence, RI.

Mathias reals:

Mathias, A.R.D. (1977). *Ann. Math. Logic* **12**, 59–111.

Ellentuck, E. (1974). *J. Symb. Logic* **39**, 163–5.

Baumgartner, J. (1983). In *Surveys in Set Theory* (Mathias, A.R.D., ed.), pp. 1–59, Cambridge University Press.

Laver reals:

Laver, R. (1976). *Acta Math.* **137**, 151–69.

Grigorieff reals:

Grigorieff, S. (1971). *Ann. Math. Logic* **3**, 363–94.

4 *Product forcing*

Let P and Q be two notions of forcing. The *product* $P \times Q$ is the coordinatewise partially ordered set product of P and Q:

$$(p_1, q_1) \leqslant (p_2, q_2) \quad \text{if } p_1 \leqslant p_2 \quad \text{and} \quad q_1 \leqslant q_2$$

If G is a generic filter on $P \times Q$, and if we let

$$G_1 = \{p \in P : \exists q (p, q) \in G\}$$
$$G_2 = \{q \in Q : \exists p (p, q) \in G\}$$

then G_1 is generic on P and G_2 is generic on Q, and $G = G_1 \times G_2$. But a stronger statement is true:

4.1 Lemma

A set G is a V-generic filter on $P \times Q$ if and only if G_1 is a V-generic filter on P and G_2 is a $V[G_1]$ generic filter on Q.

Proof $V[G_1]$-generic means of course that G_2 intersects every dense subset $D \in V[G_1]$, not just all $D \in V$. The proof is an exercise in verifying the definition of genericity. □

Neither distributivity nor chain conditions are preserved by products in general. The following lemma is trivial:

4.2 Lemma

If P and Q are both κ-closed, then so is $P \times Q$. □

To handle the c.c.c., let us introduce a stronger property:

4.3 Definition

$(P, <)$ has *property* (K) if every uncountable $X \subseteq P$ has an uncountable subset $Y \subseteq X$ whose elements are pairwise compatible.

Property (K) implies c.c.c., and we have

4.4 Lemma

If P and Q both have property (K), then so does $P \times Q$.

Proof Assume that both P and Q have property (K), and let $Z \subseteq P \times Q$ be uncountable. Either (i) there is $p \in P$ and an uncountable $Y \subset Q$ such that $\{p\} \times Y \subseteq Z$, or (ii) $\exists q \in Q$ and $X \subset P$ with $X \times \{q\} \subseteq Z$, or if neither (i) nor (ii) hold then (iii) there is an uncountable set of pairs $F \subset Z$ that is a one-to-one function. In all three cases we can use (K) for both P and Q to get an uncountable $S \subset Z$, pairwise compatible in both coordinates. □

4.5 Infinite products

Let $\{P_i : i \in I\}$ be a collection of partially ordered sets. We assume that each P_i has a greatest element, which we denote by 1.

Definition The *product* $P = \prod_{i \in I} P_i$ consists of all functions p on I with values $p(i) \in P_i$ such that $p(i) = 1$ for all but finitely many $i \in I$. The ordering of P is coordinate-wise:

$$p \leqslant q \quad \text{if } p(i) \leqslant q(i) \quad \text{for all } i \in I$$

For each $p \in P$, the finite set

$$s(p) = \{i \in I : p(i) \neq 1\}$$

is called the *support* of p. Hence, P consists of all functions in the cartesian product of P_i with finite support.

If G is a generic filter on the product $P = \prod_i P_i$, then for each $i \in I$ the set

$$G_i = \{p(i) : p \in G\}$$

is a generic filter on P_i. As in Lemma 4.1, G_i is generic not just over V, but over $V[G \restriction (I - \{i\})]$, where for every set $A \subseteq I$,

$$G \restriction A = \{p \restriction A : p \in G\}$$

The following combinatorial result is useful in computing the chain condition of a product forcing. A Δ-*system* is a collection $\mathscr{X}$ of finite sets such that

$$\text{there is a set } A \text{, the } \textit{root} \text{ of } \mathscr{X}\text{, such that for any distinct } X, Y \in \mathscr{X},\ X \cap Y = A \tag{4.6}$$

4.7 Lemma

Every uncountable set W of finite sets contains an uncountable Δ-system.

Proof Can be found in textbooks. □

Here is a typical use of a Δ-system:

4.8 Theorem

If for every $i \in I$, P_i has property (*K*), then $\prod_i P_i$ has property (*K*).

Corollary If for every $i \in I$, P_i is countable, then $\prod_i P_i$ has property (*K*).

Proof Let X be an uncountable subset of P. Let $W = \{s(p) : p \in X\}$. If W is countable, then there is a finite set $A \subset I$ such that $s(p) = A$ for uncountably many p. By Lemma 4.4, $\prod_A P_i$ has property (*K*) and the theorem follows. Thus, assume that W is uncountable. When we apply Lemma 4.7 to W, we obtain an uncountable $Z \subset X$ and a finite set $A \subseteq I$ such that $s(p) \cap s(q) = A$ whenever $p, q \in Z$, $p \neq q$. If $p, q \in Z$ and $(p \restriction A) | (q \restriction A)$ then p and q are compatible, but because $\prod_{i \in A} P_i$ has property (*K*), Z has an uncountable subset Y such that $(p \restriction A) | (q \restriction A)$ for all $p, q \in Y$. Hence Y is an uncountable subset of X of pairwise compatible elements. □

4.9 κ-products

Let κ be a regular cardinal, and let $\{P_i : i \in I\}$ be a collection of partially ordered sets. The set of all functions on I, $p(i) \in P_i$, with support $|s(p)| < \kappa$, is called the *κ-product* of the P_i. The ordering is coordinate-wise.

The following lemma is immediate:

4.10 Lemma

Let $\lambda < \kappa$. If each P_i is λ-closed then the κ-product is λ-closed. □

The Δ-system argument generalizes as follows:

4.11 Lemma

Assume that $\kappa^{<\kappa} = \kappa$. Every collection W of size κ^+ of sets of size $< \kappa$ contains a Δ-system $Z \subset W$ of size κ^+.

4.12 Theorem

Assume $\kappa^{<\kappa} = \kappa$. If for every $i \in I$, $|P_i| \leqslant \kappa$, then the κ-product of the P_i satisfies the κ^+-chain condition. □

In the particular case when $\kappa = \aleph_1$, the κ-product is the product with countable supports. In that case we shall call it the *σ-product*. Infinite products can be generalized further. For example, let $P_i, i \in I$, be partial orderings, and let J be some ideal on the index set I. Then we define

$$\prod_J P_i$$

as the set of all functions with the property that $s(p) \in J$.

Such an example is the *Easton product*. The forcing notions are indexed by ordinals, and one considers functions with *Easton support*, i.e.

$$|s(p)\cap\gamma| < \gamma \text{ for all regular cardinals } \gamma \tag{4.13}$$

Reference

For Easton forcing, see

Easton, W. (1970) *Ann. Math. Logic* **1**, 139–78.

5 Examples of product forcing

Our first example is the finite support product of many copies of Cohen forcing. This is the original Cohen's proof of independence of the continuum hypothesis.

Let κ be an infinite cardinal, and for each $i<\kappa$, let P_i be the Cohen forcing from 3.1. We let

$$P=\prod_{i<\kappa} P_i$$

be the product (with finite support) of κ copies of Cohen forcing.

The corresponding generic extension is obtained by adjoining the κ Cohen reals a_i

$$V[G]=V[\langle a_i:i<\kappa\rangle]$$

where

$$a_i=\bigcup G_i,\ G_i=\{p(i):p\in G\}$$

It is easy to see that when $i\neq j$ then every condition p forces that $\mathring{a}_i\neq\mathring{a}_j$, and so the a_i are κ distinct reals in $V[G]$.

By Theorem 4.8, P has the countable chain condition. So all cardinals are preserved in $V[G]$. Also, since $|P|=\kappa$, the size of $2^{\aleph_0}$ in $V[G]$ is at most $\kappa^{\aleph_0}$. But because there are at least κ reals in $V[G]$, the continuum in $V[G]$ has size exactly $\kappa^{\aleph_0}$. This proves the independence of the CH:

5.1 Theorem (Cohen)

For any κ such that $\kappa^{\aleph_0}=\kappa$ there is a generic extension in which $2^{\aleph_0}=\kappa$. □

Cohen forcing generalizes to regular uncountable cardinals. Let κ be a regular uncountable cardinal and let us assume GCH or at least $\kappa^{<\kappa}=\kappa$. Let P be the notion of forcing consisting of 0–1-sequences of length $<\kappa$:

$$P=\bigcup_{\alpha<\kappa}\{0,1\}^{\alpha} \tag{5.2}$$

q is stronger than p if $q\supseteq p$.

This notion of forcing adds a new subset of κ, namely

$$a = \{\alpha < \kappa : \exists p \in G\, p(\alpha) = 1\}$$

where G is the generic filter. The forcing adds no subsets of cardinals below κ, in fact no sequences of length $< \kappa$. This is because P is γ-closed for all $\gamma < \kappa$, as can easily be seen.

Under the assumption $\kappa^{<\kappa} = \kappa$, P has size κ and so it satisfies the κ^+-chain condition. It follows that all cardinals are preserved in $V[G]$: those $\leqslant \kappa$ because P is $<\kappa$-closed, and those $> \kappa$ because P has the κ^+-chain condition.

Now let $\lambda > \kappa$ be a regular cardinal (or assume that $\lambda^\kappa = \lambda$), and let, for each $i < \lambda$, P_i be the forcing notion (5.2). Let P be the κ-product of the $P_i, i \subset \lambda$:

$$P = \prod_{\substack{i<\lambda \\ <\kappa-\text{support}}} P_i \tag{5.3}$$

It is clear that P is $<\kappa$-closed. By Theorem 4.12 (we are still assuming that $\kappa^{<\kappa} = \kappa$), P satisfies the κ^+-chain condition. Hence, P preserves all cardinals.

P adjoins no new subsets of cardinals below κ. By an argument analogous to Theorem 5.1, we find that in $V[G]$

$$2^\kappa = \lambda$$

5.4 Easton's model

Cohen's independence proof admits a generalization whereby for every regular cardinal κ, the size of 2^κ in the generic extension is a prescribed cardinal λ, subject to obvious conditions on the function 2^κ, as well as the conditions required by König's theorem. (The problem of 2^κ for singular cardinals is a different story altogether.)

The model (Easton's model) is obtained by using the forcing (5.2) for each regular κ, and then taking the Easton product (4.13) of these forcings.

We shall now investigate products of the forcing notions discussed in Chapter 3.

5.5 Finite support product of random forcing

Consider adding a large number of random reals using product forcing (with finite support).

5.6 Lemma

Random forcing has property (K)

Proof Let W be an uncountable set of Borel sets in $[0,1]$ of positive measure; we shall show that uncountably many of them are pairwise compatible. We note that for some $n>0$, uncountably many $p \in W$ have measure $\geqslant 1/n$; thus assume that all $p \in W$ do and let $n>0$ be such.

The proof is by induction on n. First let $n=2$. All $p \in W$ have measure $\geqslant 1/2$. Let $W=\{p_\alpha : \alpha<\omega_1\}$. We consider $q_0=p_0$.

There is only one set of measure $\geqslant 1/2$ incompatible with q_0, namely $-q_0$. Thus $\exists \alpha_1$ such that $\forall \alpha \geqslant \alpha_1$, $p_\alpha | q_0$. Let $q_1=p_{\alpha_1}$.

Similarly, there is α_2 such that $\forall \alpha \geqslant \alpha_2$, $p_\alpha | q_0$ and $p_\alpha | q_1$; we let $q_2=p_{\alpha_2}$. We continue in this fashion and construct an uncountable set $\{q_\nu : \nu<\omega_1\} \subset W$ of pairwise compatible conditions.

Now consider $n+1$. Each $p \in W$ has measure $\geqslant 1/(n+1)$. Let $q_0=p_0$. First assume that there is an uncountable $Z \subset W$ such that every $p \in Z$ is disjoint from p_0. Z is an uncountable set of subsets of $-p_0$ (whose measure is $\leqslant n/(n+1)$), and each p has measure $\geqslant (1/n)$-times the measure of $-p_0$. By the induction hypothesis, Z has a pairwise compatible uncountable subset and we are done.

Thus, we assume that $\exists \alpha_1$ such that $\forall \alpha \geqslant \alpha_1$, $p_\alpha | q_0$. Let $q_1=p_{\alpha_1}$. Now either there is an uncountable $Z \subset W$ such that $\forall p \in Z$, $p \perp q_1$, and as before, we are done; or else $\exists \alpha_2$ such that $\forall \alpha \geqslant \alpha_2$, $p_\alpha | q_1$, and we let $q_2=p_{\alpha_2}$.

And so on: we construct an uncountable pairwise compatible set $\{q_\nu : \nu<\omega_1\}$. □

5.7 Corollary

The product of any number of copies of random forcing has the countable chain condition. □

For the next example, we consider Borel subsets of the Cantor space 2^ω, endowed with the product measure

$$m(\{x \in 2^\omega : x \supseteq \langle s(0), \ldots, s(n-1)\rangle\}) = \frac{1}{2^n}$$

For $x, y \in 2^\omega$, let $x+y=\langle x_n+y_n \bmod 2 : n<\omega\rangle$; for $A, B \subseteq 2^\omega$, let $A+B=\{a+b : a \in A, b \in B\}$.

5.8 Lemma

If $A, B \subseteq 2^\omega$ have positive measure, then $A+B$ contains an open set.

Proof There are basic open intervals I, J of the same length $l(I)=l(J)$ such that $m(A \cap I) \geqslant (2/3)l(I)$ and $m(B \cap J) \geqslant (2/3)l(J)$. We prove only the special case when $I=\{x : x(0)=0\}$, $J=\{y : y(0)=1\}$. Thus assume $A \subseteq I$, $B \subseteq J$, $m(A) \geqslant 1/3$, $m(B) \geqslant 1/3$.

We show that $A+B=J$. Let $z\in J$. The set $z-B$ is a subset of I and has measure $\geqslant 1/3$, hence it meets A and so there exist $a\in A$ and $b\in B$ such that $a+b=z$. □

5.9 Proposition

The product of (at least) two copies of random forcing adjoins a Cohen real.

Proof Let Q be the product of two random forcings, conditions $q\in Q$ are pairs of Borel sets $A,B\subseteq 2^\omega$ of positive measure. Let P be the notion of forcing consisting of nonempty open sets in 2^ω with $p_1\leqslant p_2$ iff $p_1\subseteq p_2$. P has a countable dense set (rational intervals), thus it is equivalent to Cohen forcing. We shall prove that $V^P\subset V^Q$, using Lemma 2.7.

For every $q=(A,B)\in Q$ let $h(q)$ be the interior of $A+B$ (nonempty by Lemma 5.8). If $q_1\leqslant q_2$ then $h(q_1)\leqslant h(q_2)$. To verify clause (ii) of Lemma 2.7, let $q=(A,B)$ be arbitrary and let $p'\subset \text{int}(A+B)$. We shall find $q'=(A',B')$ such that $A'\subseteq A$, $B'\subseteq B$, and such that $\text{int}(A'+B')\subseteq A'+B'$ $\subseteq p'$. We may assume that for every basic open interval I, if $A\cap I$ is nonempty, then $m(A\cap I)>0$; similarly for B.

Let U be some basic interval of length $1/2^n$ such that $U\subset p'$. If I and J are basic intervals of length $1/2^n$ then either $I+J=U$ or $I+J$ and U are disjoint. Since $U\subset A+B$, U is the union of the sets $(A\cap I)+(B\cap J)$, where I and J range over basic intervals of length $1/2^n$ and $I+J=U$. Thus, there exist such I and J so that both $A\cap I$ and $B\cap J$ have positive measure. Now we let $A'=A\cap I$, $B'=B\cap J$. □

Another way of adding a large number of random reals is to use a product measure algebra. We shall discuss it in Chapter 7.

Let us turn our attention to Sacks and Prikry–Silver reals. We prove the following results for Sacks reals, but similar arguments yield the same result for Prikry–Silver forcing. First we state the following:

5.10 Proposition

The product of two perfect set forcings preserves $\aleph_1$. □

The proof is similar to the proof of Lemma 3.5, using fusion, and $\leqslant_n$ in both coordinates. Below (Theorem 5.12) we prove a more general result.

Sacks reals cannot be added by a finite support product forcing:

5.11 Proposition

The product of infinitely many perfect set forcings (with finite support) collapses $\aleph_1$.

Proof Let P be the product of ω copies of Sacks forcing, and let $\mathscr{A}$ be an uncountable almost disjoint family of infinite subsets of ω (i.e. $A \cap A'$ is finite for any distinct $A, A' \in \mathscr{A}$). We show that in $V[G]$ there is a one-to-one mapping of $\mathscr{A}$ into ω.

Let G be a generic filter on P; $V[G] = V[\langle a_n : n \subset \omega \rangle]$, where for each n, $f_n = \bigcup\{s : s \in p(n)$ for each $p \in G\}$, and $a_n = \{k : f_n(k) = 1\}$. (The $a_n, n < \omega$, are the Sacks reals.) We shall show that for every $A \in \mathscr{A}$ there is $n = n_A$ such that $a_n \subseteq A$. This correspondence is one-to-one, as $n_A \neq n_B$ whenever A and B are almost disjoint.

Let $p \in P$, and let $A \in \mathscr{A}$ be arbitrary. Let n be such that n is not in the support of p (which is a finite set). Let T be the perfect tree

$$\{s : s(k) = 0 \quad \text{for all} \quad k \notin A\}$$

and let $q \leqslant p$ be such that $q(n) = T$. Clearly, q forces $\dot{a}_n \subseteq A_n$. □

The way to adjoin a large number of mutually generic Sacks (or Prikry–Silver) reals is to use the σ-product:

5.12 Theorem

The σ-product of any number of perfect set forcings preserves $\aleph_1$.

Remarks (1) $2^{\aleph_0} = \aleph_1$, then Sacks forcing has size $\aleph_1$, and the σ-product satisfies, by Lemma 4.11, the $\aleph_2$-chain condition. Thus all cardinals are preserved.

(2) A similar theorem holds for the Prikry–Silver forcing.

Proof We prove that the σ-product P of Sacks forcings satisfies Lemma 3.5. Let $p \in P$ force $\dot{X} : \omega \to V$; we find a countable A and a condition $q \leqslant p$ such that $q \Vdash \dot{X} \subseteq A$.

We construct a sequence $p = p_0 \geqslant p_1 \geqslant p_2 \geqslant \ldots$; let S_n denote the support of p_n. Let $S = \bigcup_{n=0}^{\infty} S_n$; by a suitable enumeration we also construct a sequence of finite sets $F_0 \subseteq F_1 \subseteq F_2 \subseteq \ldots$ with $\bigcup_{n=0}^{\infty} F_n = S$. For every n, we make sure that

$$\forall i \in F_n \quad p_{n+1}(i) \leqslant_n p_n(i)$$

This guarantees that for each $i \in S$, $p_\infty(i) = \bigcap_{n=0}^{\infty} p_n(i)$ is a perfect tree, and so $p_\infty = \langle p_\infty(i) \rangle_i$ is a condition (with support S), stronger than all p_n.

The sequence $\{p_n\}_n$ is constructed by induction. At stage n, we construct p_n from p_{n-1}, and also a set A_n, and at the end we let $A = \bigcup_{n=0}^{\infty} A_n$.

Consider p_{n-1} and F_n. For each $i \in F_n$, let E_i be the set of all successors of all nth branching points of the tree $p_{n-1}(i)$ (the set E_i has 2^{n+1} elements). Let $\sigma_1, \ldots, \sigma_l$ be all the functions σ on F_n such that $\sigma(i) \in E_i$, $\forall i \in F_n$. We let $q_0 = p_{n-1}$, and then construct $q_0 \geqslant q_1 \geqslant \cdots \geqslant q_l$ and $A_n = \{a_1, \ldots, a_l\}$ as follows: given q_k, consider σ_k. Find $r \leqslant q_k$ such that $r(i) \leqslant q_k(i) \restriction \sigma_k(i)$, and a_k such that

$$r \Vdash \dot{X}(n) = a_k \tag{5.13}$$

Then amalgamate r into q_k to get q_{k+1}: $q_{k+1}(i)$ is the amalgamation of $r(i)$ into $q_k(i)$ if $i \in F_n$, and is $r(i)$ for all other i. It follows that $q_{k+1}(i) \leqslant_n q_k(i)$ for all $i \in F_n$. Finally, let $p_n = q_l$.

It remains to prove that $p_\infty \Vdash \dot{X} \subseteq A$. So let $q \leqslant p_\infty$ and $n \in \omega$. There is $\bar{q} \leqslant q$ and some a such that $\bar{q} \Vdash \dot{X}(n) = a$. Consider the set F_n, and the corresponding $\sigma_1, \ldots, \sigma_l$. There is some $k \leqslant l$ such that $\sigma_k(i) \in \bar{q}(i)$ for all $i \in F_n$, and therefore $\bar{q} \restriction \sigma_k$ is stronger than the r in (5.13). It follows that $a \in A_n$, and so $\bar{q} \Vdash \dot{X}(n) \in A$. Hence, $p_\infty \Vdash \dot{X} \subseteq A$. □

Next we consider products of Laver and Mathias reals.

5.14 Proposition

The product of two Laver (or Mathias) forcings preserves $\aleph_1$.

Proof A suitable modification of Lemma 3.20 or Lemma 3.12. For instance, consider Laver reals. As in Lemma 3.20, we are given $(p, p') \Vdash \dot{X}: \omega \to A$, and want to find $(q, q') \leqslant (p, p')$ and a countable A such that $(q, q') \Vdash \dot{X} \subseteq A$. We construct a sequence (p_n, p'_n) such that $p_{n+1} \leqslant_n p_n$ and $p'_{n+1} \leqslant_n p'_n$ for all n; we also construct $A = \bigcup_{n=0}^{\infty} A_n$.

At stage n, we run through all *pairs* (t_i, t'_i) of maximal nodes among $\{s_0^{p_n}, \ldots, s_n^{p_n}\} \times \{s_0^{p'_n}, \ldots, s_n^{p'_n}\}$ and successively amalgamate those (q_i, q'_i) with stem (t_i, t'_i) (into (q_{i-1}, q'_{i-1})) that decide the value of $\dot{X}(u_n)$.

The proof that $(p_\infty, p'_\infty) \Vdash \dot{X} \subseteq A$ is similar to the proof of Lemma 3.20. □

The following is worth mentioning.

5.15 Proposition

The product of two copies of Laver forcing (or of Mathias forcing) adjoins a Cohen real.

Proof The product of two Laver forcings has a dense set Q such that if $q = (T, T') \in Q$ then T and T' have stems of equal length. Let P be the Cohen forcing. We shall prove that $V^P \subset V^Q$, using Lemma 2.7.

For every $q \in Q$ with stems s, s' let $h(q) = p$ be the Cohen condition defined by

$$p(k) = \begin{cases} 1 & \text{if} \quad s(k) \leqslant s'(k) \\ 0 & \text{if} \quad s(k) > s'(k) \end{cases} \qquad k < \text{length}\,(s)$$

If $q_1 \leqslant q_2$ then $h(q_1) \leqslant h(q_2)$. To verify clause (ii) of Lemma 2.7, let q be arbitrary, and let $p' \supseteq h(q)$. It is easy to find $q' \leqslant q$ such that $h(q') = p'$. Thus, $V^P \subseteq V^Q$. □

As for Sacks reals, Laver and Mathias reals cannot be added by a finite support product forcing:

5.16 Proposition

The product of infinitely many Laver forcings or Mathias forcings (with finite support) collapses $\aleph_1$.

Proof Exactly the same as the proof of Proposition 5.11. The key fact is that for any infinite set A there is a condition p that forces $\dot{a} \subseteq A$, where $\dot{a}$ is either a Laver or Mathias real. For Laver forcing, we can take

$$p = \{t\colon t(n) \in A \text{ for every } n \in \text{dom}\,(t)\};$$

in Mathias forcing, we take

$$p = (\varnothing, A) \quad \square$$

But σ-product does not work either, if one wants to add many Laver or Mathias reals:

5.17 Proposition (Baumgartner)

The σ-product of infinitely many Laver forcings (or Mathias forcings) collapses $2^{\aleph_0}$.

Proof We give the proof for Mathias reals, and only under the assumption of $2^{\aleph_0} = \aleph_1$; the argument is slightly more complicated if CH is not assumed. Let P be the σ-product of ω copies of Mathias forcing. A condition is a sequence $\{(s_n, S_n)\colon n \subset \omega\}$, where each (s_n, S_n) is a Mathias condition. A generic filter G yields a sequence $\{f_n\}_n$ of Mathias reals, $f_n\colon \omega \to \omega$.

Using the Mathias reals f_n, we define, for each k, a function $h_k\colon \omega \to \omega$ as follows:

$$h_k(0) = f_0(k), \quad h_k(n+1) = f_{n+1}(h_k(n))$$

i.e.

$$h_k(n) = f_n f_{n-1} \cdots f_1 f_0(k)$$

Let $H = \{h_k : k \in \omega\}$. In the ground model, we assume the CH and so $\omega^\omega = \{e_\alpha : \alpha < \omega_1\}$. It suffices to show that in $V[G]$, the countable set $\{\alpha : e_\alpha \in H\}$ is cofinal in ω_1^V.

Let $C \subset \omega^\omega$ be a countable set in V. It is enough to find an $h \in H$ such that $h \in V$ and $h \notin C$. Let $p = \{(s_n, S_n)\}_n$ be a condition. We shall find $k \in \omega$, $q = \{(t_n, T_n)\}_n \leqslant p$ and $h \in \omega^\omega$ such that $h \notin C$ and $q \Vdash \dot{h}_k = h$. Let $C = \{g_n : n \in \omega\}$.

First we find $k > |s_0|$ and $(t_0, T_0) < (s_0, S_0)$ so that $t_0(k) >$ both $|s_1|$ and $g_0(0)$, and let $h(0) = t_0(k)$. Then we find $(t_1, T_1) < (s_1, S_1)$ so that $t_1(h(0)) >$ both $|s_2|$ and $g_1(1)$, and let $h(1) = t_1(h(0))$. In general, we find $(t_n, T_n) < (s_n, S_n)$ so that

$$h(n) = t_n(t_{n-1} \ldots (t_0(k)) \ldots) > \text{both } |s_{n+1}| \quad \text{and} \quad g_n(n)$$

It follows that $h \neq g_n$ for all n, and that the condition $\{(t_n, T_n)\}_n$ forces $\dot{h}_k = h$. □

It is possible to add many mutually generic Laver or Mathias reals, but one has to use a different kind of product; one such method (product with *mixed support*) will be discussed in Part III, Chapter 8.

We conclude this chapter by stating, without proof, the relevant result on products of Grigorieff reals:

5.18 Proposition

The σ-product of any number of Grigorieff forcings preserves $\aleph_1$. □

6 *The Lévy collapse*

When λ is any cardinal, it is easy to make λ countable in a generic extension: let P be the set of all finite sequences

$$p = \langle \alpha_0, \alpha_1, \ldots, \alpha_{n-1} \rangle$$

of ordinals $< \lambda$; $q \leqslant p$ means that q extends p.

A generic filter yields a function $f = \bigcup G$; it is easy to see that f maps ω onto λ. (For every α, the set $\{p : \alpha \in \text{range}\,(p)\}$ is dense.) A chain condition argument shows that cardinals λ^+ and up are preserved.

More generally, when κ is regular, and $\lambda > \kappa$ is such that $\lambda^{<\kappa} = \lambda$, the forcing with sequences of length $< \kappa$ of ordinals $< \lambda$ collapses λ onto κ (i.e. $|\lambda| = \kappa$ in $V[G]$) and preserves all cardinals $\leqslant \kappa$ and $> \lambda$. This is because the forcing is $< \kappa$-closed and has size λ.

The *Lévy collapse* is a forcing notion that collapses cardinals below a given inaccessible cardinal κ while preserving κ. The prototypical case is when all cardinals below κ are made countable.

6.1 Theorem (Lévy)

Let κ be an inaccessible cardinal. There is a forcing P such that in V^P, κ becomes $\aleph_1$.

Proof Let P be the set of all functions p from a finite subset of $\kappa \times \omega$ into κ, such that

$$p(\alpha, n) < \alpha$$

for all $\alpha < \kappa$ and all n; let

$$p \leqslant q \quad \text{if} \quad p \text{ extends } q$$

If G is a generic filter on P, let $f = \bigcup G$, and let for each $\alpha < \kappa$,

$$f_\alpha(n) = f(\alpha, n)$$

For each $\alpha < \kappa$, the function f_α maps ω onto α; this is because for each $\beta < \alpha$, the set $\{p : \exists n\, p(\alpha, n) = \beta\}$ is dense. Thus, all cardinals below κ become countable in $V[G]$. In order to show that κ is preserved, it is enough to verify the chain condition:

6.2 Lemma

$(P, <)$ satisfies the κ-chain condition.

Proof By a Δ-system argument. Let $\{p_\alpha : \alpha < \kappa\}$ be conditions. For each α, let $p_\alpha^- = p_\alpha \upharpoonright (\alpha \times \omega)$ and $p_\alpha^+ = p - p_\alpha^-$. By Fodor's theorem, there is a stationary $S \subseteq \kappa$ and some q such that $p_\alpha^- = q$ for all $\alpha \in S$. Now we construct an increasing sequence α_ν, $\nu < \kappa$, in S so that whenever $\xi < \eta$, then $\mathrm{dom}(p_{\alpha_\xi}^+) \subset \alpha_\eta \times \omega$. It follows that any two p_{α_ξ} and p_{α_η} are compatible. □

The theorem generalizes as follows: let λ be regular and let $\kappa > \lambda$ be inaccessible. The forcing conditions are functions from a subset of $\kappa \times \lambda$ of size $< \lambda$, into κ, and such that

$$p(\alpha, \beta) < \alpha$$

for all $\alpha < \kappa$ and all $\beta < \lambda$. The resulting model collapses all cardinals α such that $\lambda \leqslant \alpha < \kappa$ onto λ, and makes κ the successor of λ. All cardinals below λ are preserved, as the forcing is $< \lambda$-closed. The forcing 6.1 (and its generalization) is called the Lévy collapse.

An important feature of the Lévy collapse is its homogeneity. Let B be the complete Boolean algebra $B(P)$ for the forcing 6.1. It can be shown that B is homogeneous, i.e. for any $u, v \in B$ other than 0 or 1 there is an automorphism π of B such that $\pi u = v$. Invoking Proposition 2.2 we note that no real number in $V[G]$ is definable with parameters in V. Therefore there is no well ordering of $\mathbf{R}$ definable over V.

In fact, the Lévy algebra satisfies this strong version of homogeneity (which we state without proof):

6.3 Theorem

If A_1 and A_2 are isomorphic complete subalgebras of B of size $< \kappa$, and if π is an isomorphism between A_1 and A_2, then π can be extended to an automorphism of B. □

This homogeneity is used to prove the following theorem:

6.4 Theorem

In the Lévy model, there is no well ordering of $\mathbf{R}$ in $L(\mathbf{R})$. □

(In particular, there is no projective well ordering of the reals.) And most notably:

6.5 Theorem (Solovay)

In the Lévy model, all sets of reals in $L(\mathbf{R})$ are Lebesgue measurable. □

And the submodel $L(\mathbf{R})$ of the Lévy model is the celebrated model (without AC) in which all sets of reals are Lebesgue measurable.

References

Solovay, R. (1970). *Ann. Math.* **92**, 1–56.
Jech, T. (1978). *Set Theory*, Sections 25, 42, Academic Press, NY.

7 Product measure forcing

In this chapter we describe a method of adding a large number of random reals by means of forcing with measure algebras.

In 3.2 we used the measure algebra of Borel sets in R modulo Lebesgue null sets to add a random real. We shall now concern ourselves with more general measure algebras.

Let Ω be a set, let $\mathscr{S}$ be a σ-algebra of subsets of Ω, and let m be a σ-additive probability measure defined on $\mathscr{S}$; in particular, $m(\emptyset)=0$ and $m(\Omega)=1$. Let B be the Boolean algebra $\mathscr{S}/I$, where I is the ideal of all $X\in\mathscr{S}$ of measure 0. B is a complete Boolean algebra.

The measure m induces a measure on B; also denoted by m:

$$m(X/I)=m(X)$$

and we have $m(0)=0$, $m(1)=1$ and $m(a)>0$ whenever $a\in B$ is not zero. We call B a *measure algebra* and (Ω, m) the corresponding *measure space*. We shall not distinguish between elements of B and corresponding sets in $\mathscr{S}$.

Every measure algebra B satisfies the c.c.c. and so forcing with B preserves cardinals. The size of the continuum in V^B is related to the size of B. The following lemma gives a representation of real numbers in V^B:

7.1 Lemma (Scott)

There is a one-to-one correspondence between real numbers in V^B and (real-valued) measurable functions on Ω (mod a.e.).

Proof We shall not give the proof but state explicitly the correspondence. With each (name for a) real $\dot{r}$ in V^B one associates a measurable function $f_r:\Omega\to R$ such that for each $\lambda\in R$,

$$\|\dot{r}\leqslant\lambda\|=\{x\in\Omega: f_r(x)\leqslant\lambda\} \tag{7.2}$$

The one-to-one correspondence means that for any $\dot{r}$ and $\dot{s}$,

$$\|\dot{r}=\dot{s}\|=\{x\in\Omega: f_r(x)=f_s(x)\}$$

Moreover, $+,\cdot$ and $\leqslant$ are preserved by the correspondence. □

To add a large number of random reals, we use *product measure* algebra. Let I be an infinite set and let

$$\Omega = \Omega_I = \{0, 1\}^I$$

Let T be the set of all functions t from a finite subset of κ into $\{0, 1\}$. For each $t \in T$, let

$$S_t = \{f \in \Omega : t \subset f\}$$

and let $\mathcal{S} = \mathcal{S}_I$ be σ-algebra σ-generated by $\{S_t : t \in T\}$. The *product measure* $m = m_I$ on $\mathcal{S}$ is the unique measure so that

$$m(S_t) = \frac{1}{2^{|t|}}$$

Let $B = B_I$ be the corresponding measure algebra.

If G is a generic ultrafilter on B, then

$$f = f_G = \bigcup \{t : S_t \in G\} \tag{7.3}$$

is a function in $V[G]$, $f : I \to \{0, 1\}$, and $V[G] = V[f]$.

Let κ be an infinite cardinal and let $I = \kappa \times \omega$. Let $B = B_I$, and let f: $\kappa \times \omega \to \{0, 1\}$ be as in (7.3). Clearly, B_I is isomorphic to B_κ. For each $\alpha < \kappa$, we let $f_\alpha \in \{0, 1\}^\omega$ be the function

$$f_\alpha(n) = f(\alpha, n)$$

The f_α, $\alpha < \kappa$, are κ distinct random reals, and the continuum in $V[G]$ has size $\geqslant \kappa$. (Because $(2^{\aleph_0})^{V[G]} \leqslant |B|^{\aleph_0} = \kappa^{\aleph_0}$, we have $(2^{\aleph_0})^{V[G]} = \kappa^{\aleph_0}$.)

One application of forcing with a product measure algebra is Solovay's model for a real valued measurable cardinal.

7.4 Definition

(*a*) An uncountable cardinal κ is *measurable* if there exists a κ-complete nonprincipal ultrafilter over κ.

(*b*) An uncountable cardinal κ is *real-valued-measurable* if there exists a nontrivial κ-additive measure on κ.

7.5 Theorem (Solovay)

If κ is a measurable cardinal then adjoining κ random reals (i.e. forcing with the product measure algebra B_κ) makes κ a real-valued-measurable cardinal (and $2^{\aleph_0} = \kappa$ in V^B).

The theorem follows from Lemma 7.6. If κ is a measurable cardinal and U is a κ-complete nonprincipal ultrafilter on κ, then $\mu = \mu_U$ defined by

$$\mu(X)=\begin{cases}1 & \text{if } X\in U\\ 0 & \text{if } X\notin U\end{cases}$$

is a nontrivial κ-additive measure on κ.

7.6 Lemma

Let U be a κ-complete nonprincipal ultrafilter over κ and let B be a measure algebra. In $V[G]$, there is a κ-additive measure ν on κ extending μ_U.

Proof We give the definition of the extending measure without verifying that the definition works, as the detailed proof is rather tedious.

Let m be the measure on B and let G be a generic ultrafilter on B. If $A\subseteq\kappa$ in $B[G]$, we let $\dot{A}$ be a name for A and define

$$\nu(A)=\lim_{a\in G}\lim{}_U\frac{m(a\cdot\|\alpha\in\dot{A}\|)}{m(a)} \tag{7.7}$$

The limit $\lim_U$ of the expression in (7.7) is the unique real number that the expression takes for U-almost all $\alpha\in\kappa$.

One still has to show that the limit $\lim_G$ exists, that the value $\nu(A)$ is independent of the choice of the name $\dot{A}$, and that ν is a κ-additive measure. □

An extension of Solovay's Theorem 7.5 has been used in point-set topology in connection with the *normal Moore space conjecture.*

We recall the definition of strongly compact cardinals:

7.8 Definition

If A is a set, $|A|\geqslant\kappa$, we let

$$P_\kappa(A)=[A]^{<\kappa}=\{x\subset A:|x|<\kappa\}$$

An uncountable cardinal κ is *strongly compact* if for every $\lambda\geqslant\kappa$ there exists a *fine* ultrafilter on $P_\kappa(\lambda)$, i.e. a κ-complete ultrafilter U such that for every $z\in P_\kappa(\lambda)$,

$$\{x\in P_\kappa(\lambda):z\subseteq x\}\in U$$

7.9 Lemma

If U is a fine ultrafilter on $P_\kappa(\lambda)$, there are functions $h_\alpha:P_\kappa(\lambda)\to\kappa$, $\alpha<\lambda$, such that $h_\alpha<h_\beta \bmod U$ whenever $\alpha<\beta$.

Proof For each α, let

$$h_\alpha(x) = \text{order type of } x \cap \alpha \quad \square$$

7.10 Theorem (Kunen)

If κ is a strongly compact cardinal then the model $V[G]$ obtained by adding κ random reals satisfies

$$\text{for every } \lambda \geqslant \kappa, \text{ the product measure } m_\lambda \text{ on } \{0,1\}^\lambda \text{ can be extended to a } 2^{\aleph_0}\text{-additive measure defined on all subsets of } \{0,1\}^\lambda \tag{7.11}$$

Proof Let U be a fine ultrafilter on the set $\Gamma = P_\kappa(\lambda)$ (in V). Solovay's construction in Theorem 7.5 extends μ_U to a κ-additive measure ν on Γ (in $V[G]$). Let $f_G: \kappa \to \{0,1\}$ be the function defined in (7.3).

Using f_G and the functions h_α from Lemma 7.9, we define a function $F: \Gamma \to \{0,1\}^\lambda$ as follows:

$$(F(x))(\alpha) = f_G(h_\alpha(x)) \quad x \in \Gamma, \alpha \in \lambda$$

and define a κ-additive measure σ on $\{0,1\}^\lambda$ by

$$\sigma(X) = \nu(F^{-1}(X))$$

It is not difficult to verify that σ extends the product measure m_λ on $\{0,1\}^\lambda$. $\square$

Property (7.11) is instrumental in Nyikos' consistency proof of the normal Moore space conjecture:

7.12 Theorem (Nyikos)

If (7.11) holds then every normal Moore space is metrizable. $\square$

References

Solovay, R. (1971). In *Axiomatic Set Theory* (Scott, D., ed.), pp. 397–428, American Math. Society, Providence, RI.

Fleissner, W. (1984). In *Handbook of Set-theoretic Topology* (Kunen, K. and Vaughan, J.E., eds.), pp. 733–60, North-Holland, Amsterdam.

Jech, T. (1978). *Set Theory*, Section 34, Academic Press, N.Y.

[illegible] measure [illegible]

Proof. For each [illegible]

[illegible] order type [illegible]

7.10 [illegible]

[illegible] strongly [illegible] that [illegible] 1 [illegible]

[illegible] ing [illegible] is [illegible]

[illegible] every [illegible] the product [illegible] on $[0,1]^2$ [illegible] be extended to a 2-additive measure defined on [illegible] subsets of $[0,1]$ [illegible]

Proof. Let U be a fine ultrafilter on the [illegible] construction in Theorem [illegible] a 2-additive measure [illegible]

Using [illegible] and the functions [illegible] define a function [illegible] $[0,1]$ as follows:

[illegible]

and define a 2-additive measure ν on $[0,1]$ [illegible]

[illegible]

It is not difficult to [illegible] that ν extends the product measure on $[0,1]^2$.

[illegible] is [illegible] in [illegible] proof of the [illegible] space [illegible]

7.12 Theorem [illegible]

[illegible] approximation

References

[illegible]

[illegible]

[illegible]

PART II

Iterated forcing

1 *Two step iteration*

Let P be a notion of forcing, and let $\dot{Q} \in V^P$ be a name for a partial ordering in V^P. There is a notion of forcing $P*\dot{Q}$ (in V) with the property that forcing with $P*\dot{Q}$ amounts to the same as first forcing with P and then (in the extension by P) with $\dot{Q}$. Let

$$\begin{aligned} &P*\dot{Q} = \{(p,\dot{q}) \colon p \in P \quad \text{and} \quad \|q \in \dot{Q}\|_P = 1\} \\ &(p_1,\dot{q}_1) \leqslant (p_2,\dot{q}_2) \quad \text{iff} \quad p_1 \leqslant p_2 \quad \text{and} \quad p_1 \Vdash \dot{q}_1 \leqslant \dot{q}_2 \end{aligned} \tag{1.1}$$

(1.1) defines a partially ordered set $P*\dot{Q}$, the *two step iteration* of P and $\dot{Q}$.

1.2 Lemma

Let G be a V-generic filter on P and let $Q = \dot{Q}/G$ be a notion of forcing in $V[G]$, with a name $\dot{Q} \in V^P$. If H is $V[G]$-generic on Q, then

$$G*H = \{(p,\dot{q}) \in P*\dot{Q} \colon p \in G \quad \text{and} \quad \dot{q}/G \in H\}$$

is a V-generic filter on $P*\dot{Q}$ and $V[G*H] = V[G][H]$. ($\dot{Q}/G$ and $\dot{q}/G$ denote the G-interpretations of the names $\dot{Q}$ and $\dot{q}$.)

Proof Let us prove that $G*H$ meets every dense subset of $P*\dot{Q}$. If D is dense, let (in $V[G]$)

$$D_1 = \{\dot{q}/G \colon \exists p \in G \quad \text{such that} \quad (p,\dot{q}) \in D\}$$

The set D_1 is dense in $Q = \dot{Q}/G$: this is proved by showing that for every $\dot{q}_0$, the set (in V)

$$\{p \in P \colon \exists \dot{q}_1 (p \Vdash \dot{q}_1 \leqslant \dot{q}_0 \quad \text{and} \quad (p,\dot{q}_1) \in D)\}$$

is dense in P. Hence, D_1 meets H and so there is $q \in H$ such that for some $p \in G$ and some $\dot{q}$, $q = \dot{q}/G$ and $(p,\dot{q}) \in D$. It follows that $(p,\dot{q}) \in D \cap (G*H)$. □

1.3 Lemma

Let K be V-generic on $P*\dot{Q}$. The set

$$G = \{p \in P \colon \exists \dot{q} (p,\dot{q}) \in K\}$$

is a V-generic filter on P and

$$H = \{\dot{q}/G : \exists p (p, \dot{q}) \in K\}$$

is a $V[G]$-generic filter on $Q = \dot{Q}/G$. Moreover, $K = G * H$.

Proof (*a*) Let D (in V) be dense in P. Then $D_1 = \{(p, \dot{q}) : p \in D\}$ is dense in $P * \dot{Q}$ and it follows that $D \cap G$ is nonempty.

(*b*) Let $D \in V[G]$ be dense in Q, and let $\dot{D} \in V^P$ be a name for D such that $\Vdash_P \dot{D}$ is dense in $\dot{Q}$. Then

$$\{(p, \dot{q}) \in P * \dot{Q} : p \Vdash \dot{q} \in \dot{D}\}$$

is dense in $P * \dot{Q}$ and it follows that $H \cap D$ is nonempty. □

The two lemmas establish that

$$V[G * H] = V[G][H]. \tag{1.4}$$

We shall now describe the complete Boolean algebra $B(P * \dot{Q})$. Let $B = B(P)$, and let $\dot{C} \in V^B$ be such that $\| \dot{C} = B(\dot{Q}) \| = 1$. Let

$$D_0 = \{\dot{c} : \| \dot{c} \in \dot{C} \| = 1\} \tag{1.5}$$

and for all $\dot{c}_1, \dot{c}_2 \in D_0$, let

$$\dot{c}_1 \equiv \dot{c}_2 \quad \text{iff} \quad \| \dot{c}_1 = \dot{c}_2 \| = 1 \tag{1.6}$$

(1.6) is an equivalence relation on D_0; we denote the quotient of D_0 by $\equiv$

$$D = B * \dot{C} \tag{1.7}$$

D is a Boolean algebra, with Boolean-algebraic operations defined in the natural way: e.g.

$$a = a_1 + a_2 \quad \text{iff} \quad \| a = a_1 + a_2 \| = 1$$

Routine arguments show that $B * \dot{C}$ is a complete Boolean algebra, and that $B * \dot{C} = B(P * \dot{Q})$. The partial order $P * \dot{Q}$ is embedded in $B * \dot{C}$ as a dense subset as follows: $(p, \dot{q}) \mapsto \dot{c}$, where

$$\| \dot{c} = \dot{q} \| = p, \quad \| \dot{c} = 0 \| = -p$$

Since $V[G] \subseteq V[G][H] = V[G * H]$, we have

$$V^P \subseteq V^{P * \dot{Q}} \tag{1.8}$$

The following gives an explicit embedding of B into $B * \dot{C}$: for each $b \in B$, let $\dot{c} = \pi(b)$ be the unique $\dot{c}(\text{mod} \equiv)$ such that

$$\| \dot{c} = 1 \| = b \quad \text{and} \quad \| \dot{c} = 0 \| = -b$$

The mapping π is a complete embedding of B into $B*\dot{C}$. (Thus, B can be considered a complete subalgebra of $B*\dot{C}$.)

A two step iteration is in a sense a generalization of a product of two forcing notions. Let P and Q be two partial orderings in V. When we identify Q with its canonical name in V^P, we can form the partial order $P*Q$. It is easy to verify that $P\times Q$ is a dense subset of $P*Q$, and so we have $V^{P\times Q}=V^{P*Q}$.

Two step iterations preserve closure conditions and chain conditions.

1.9 Proposition

(*a*) If P is κ-closed and if $\Vdash \dot{Q}$ is κ-closed, then $P*\dot{Q}$ is κ-closed.

(*b*) If P is κ-distributive and $\Vdash \dot{Q}$ is κ-distributive then $P*\dot{Q}$ is κ-distributive.

Proof (*a*) Let $\lambda\leqslant\kappa$ and let

$$(p_1,\dot{q}_1)\geqslant(p_2,\dot{q}_2)\geqslant\cdots\geqslant(p_\alpha,\dot{q}_\alpha)\geqslant\cdots \qquad (\alpha<\lambda)$$

be a descending sequence in $P*\dot{Q}$. Then $\{p_\alpha\}_\alpha$ is a descending sequence in P and so has a lower bound p. The condition p forces that $\{\dot{q}_\alpha\}_\alpha$ is a descending sequence in $\dot{Q}$ and, therefore, that it has a lower bound $\dot{q}$. But then $(p,\dot{q})$ is a lower bound of $\{(p_\alpha,\dot{q}_\alpha)\}_\alpha$.

(*b*) If P does not add new κ-sequences, and if in V^P, $\dot{Q}$ does not add new κ-sequences, then $P*\dot{Q}$ does not add new κ-sequences. □

1.10 Theorem

(*a*) Let κ be regular. If P has the κ-chain condition and if $\Vdash \dot{Q}$ has the κ-chain condition, then $P*\dot{Q}$ has the κ-chain condition.

(*b*) If P has property (K) and $\Vdash \dot{Q}$ has property (K), then $P*\dot{Q}$ has property (K).

Proof We recall a consequence of the chain condition:

1.11 Lemma

If P has the κ-c.c. and if $\Vdash(\dot{Z}\subseteq\kappa$ and $|\dot{Z}|<\kappa)$, then for some $\gamma<\kappa$, $\Vdash \dot{Z}\subseteq\gamma$.

Proof By Theorem 2.14 in Part I, $\Vdash \kappa$ is regular. Let W be a maximal antichain such that $\forall p\in W\exists\gamma_p$ with $p\Vdash \dot{Z}\subseteq\gamma_p$. Let $\gamma=\sup\gamma_p$. □

(*a*) Assume that $(p_\alpha,\dot{q}_\alpha)$, $\alpha<\kappa$, are mutually incompatible. In $V[G]$ (for a V-generic G on P), let $Z=\{\alpha:p_\alpha\in G\}$; the canonical name for Z is such that $\|\alpha\in\dot{Z}\| = p_\alpha$. For any α and β, $p_\alpha\cdot p_\beta \Vdash \dot{q}_\alpha \perp \dot{q}_\beta$ (unless $p_\alpha \perp p_\beta$), as is easy to verify. Thus, $q_\alpha \perp q_\beta$ whenever $\alpha, \beta\in Z$. Since Q has the κ-c.c. in $V[G]$, we have $|Z| < \kappa$. By Lemma 1.11 there is $\gamma < \kappa$ such that $\Vdash \dot{Z} \subseteq \gamma$, but that contradicts the fact that $p_\gamma \Vdash \gamma\in\dot{Z}$.

(*b*) Let $(p_\alpha, \dot{q}_\alpha)$, $\alpha < \omega_1$, be an uncountable set of conditions in $P * \dot{Q}$. As in (*a*), let $Z = \{\alpha : p_\alpha\in G\}$, where G is V-generic on P. Let

$$b = \|\dot{Z} \text{ is uncountable}\|$$

Since $b = 0$ leads to a contradiction (as in (*a*)), we have $b \neq 0$. Now $\dot{Q}$ has property (*K*), and so

$$b \Vdash \exists \text{ uncountable } \dot{W} \subseteq \dot{Z} \text{ such that } \dot{q}_\alpha | \dot{q}_\beta \text{ for all } \alpha, \beta\in\dot{W}.$$

For every α, let $b_\alpha = b\cdot\|\alpha\in\dot{W}\|$; note that $b_\alpha \leqslant p_\alpha = \|\alpha\in\dot{Z}\|$. The set $\{\alpha : b_\alpha \neq 0\}$ is uncountable, and by (*K*) there is an uncountable set Y such that $b_\alpha | b_\beta$ for all $\alpha, \beta\in Y$.

Now if α and β are in Y, we have

$$b_\alpha\cdot b_\beta \Vdash \dot{q}_\alpha | \dot{q}_\beta$$

and it follows that $(p_\alpha, \dot{q}_\alpha) | (p_\beta, \dot{q}_\beta)$. □

Neither distributivity nor chain conditions are generally preserved by products. By Theorem 1.10, if P has the κ-c.c. then a product $P \times Q$ has the κ-c.c. if Q has the κ-c.c. in V^P, but that does not necessarily follow from Q having the κ-c.c. in V.

1.12 Example

Let P be the forcing that adjoints a Cohen real, and let G be generic on P. Let Q be the Cohen forcing in $V[G]$. Clearly, $Q\in V$ because its conditions are finite functions under extension; and Q is, in V, the Cohen forcing P. Thus, $V^{P*Q} = V^{P\times P}$, and the iteration amounts to forcing with $P \times P$ (which is of course isomorphic to P). Thus, adding a Cohen real and then adding another Cohen real is like adding a single Cohen real.

Example Let P be the random real forcing, and let $\dot{Q}$ be the random real forcing in V^P. We claim that $B(P * \dot{Q})$ is a (countably generated) measure algebra, and so adding a random real a and then adding a random real over $V[a]$ is like adding a single random real.

1.13 Proposition

If B is a measure algebra and if $\|\dot{C}$ is a measure algebra $\|_B = 1$, then $B * \dot{C}$ is a measure algebra.

Proof Let μ be a measure on B and let $\dot{m}$ be in V^B a measure on $\dot{C}$. Let D_0 be the set (1.5). For each $\dot{c}\in D_0, \dot{m}(\dot{c})$ is in V^B a real number between 0 and 1, and as such is represented by a measurable function in V with values in $[0,1]$ (see Solovay's paper for details). We denote this function $m(\dot{c})$. When we define

$$\nu(\dot{c}) = \int m(\dot{c})\,\mathrm{d}\mu$$

the value is preserved by the equivalence (1.6), and we obtain a measure on $B*\dot{C}$. □

1.14

The operation $B*\dot{C}$ has an inverse. Let D be a complete Boolean algebra and let B be a complete subalgebra of D. There is a complete Boolean algebra $\dot{C}$ in V^B such that $B*\dot{C}$ is isomorphic to D. In V^B, let $\dot{J}$ be the ideal on D generated by the dual of the canonical generic ultrafilter $\dot{G}$ on B:

$$\|d\in\dot{J}\|_B = \sum\{b\in B: b\cdot d = 0\} \quad (d\in D) \tag{1.15}$$

In $V_B, \dot{C}$ is the quotient algebra $D/\dot{J}$. We denote the algebra $\dot{C}$ by $D{:}B$.

1.16 Example

Let a be a Cohen real, and let $b\in V[a]$ be a real. Then either $V[a]=V[b]$ or $V[a]=V[b][c]$, where c is a Cohen real over $V[b]$. Because if D is the algebra that adjoins a, and B is a complete subalgebra of D, then $D{:}B$ has (in V^B) a countable dense subset.

1.17 Example

If a is a random real and b a real in $V[b]$ then either $V[a]=V[b]$ or $V[a]=V[b][c]$, where c is a random real over $V[b]$. More generally, let D be a measure algebra with measure ν and let $B\subseteq D$ be a complete subalgebra with measure $\mu=\nu\restriction B$. Let $\dot{C}=D{:}B$. Then $\dot{C}$ is a measure algebra in V^B.

To see this, consider the measures $\nu_d, d\in D$, on B:

$$\nu_d(b) = \frac{\nu(b\cdot d)}{\nu(d)} \qquad (b\in B)$$

By the Radon–Nikodým theorem, there are measurable functions f_d on the measure space for B such that

$$\nu_d(b) = \int f_d\,\mathrm{d}\mu$$

Each f_d represents a real number $\dot{r}_d \in V^B$. We define a measure $\dot{\mu}$ on $D/\dot{J}$ in V^B by

$$m(d/\dot{J}) = \dot{r}_d$$

1.18 Example

Let κ be inaccessible, and let $V[G]$ be the Lévy collapse. If a is a real in $V[G]$ then $V[G] = V[a][H]$, where H is a Lévy collapse. If D is the Lévy collapse algebra and $B \subseteq D$ is such that $|B| < \kappa$ then $D{:}B$ is, in V^B, the Lévy collapse algebra. This follows from the homogeneity of the Lévy algebra.

References

Solovay, R. and Tennenbaum, S. (1971). *Ann. Math.* **94**, 201–45.
Jech, T. (1978). *Set Theory*, Section 23, Academic Press, NY.
Solovay, R. (1971). Real-valued measurable cardinals, *Proc. Symp. Pure Math.* **13**, I, American Math. Society 1971, pp. 397–428.

2 Finite support iteration

We generalize the two step iteration of Chapter 1 as follows: we shall introduce sequences $\{P_\alpha\}_\alpha$ of forcing notions so that

$$P_{\beta+1} = P_\beta * \dot{Q}_\beta \tag{2.1}$$

when $\alpha = \beta + 1$, where $\dot{Q}_\beta$ is some forcing notion in V^{P_β}; if α is a limit ordinal, then

$$P_\alpha = \text{direct limit of } P_\beta, \beta < \alpha \tag{2.2}$$

To explain the meaning of (2.2), let $B_\alpha = B(P_\alpha)$. The iteration is so defined that for $\beta \leqslant \alpha$, B_β is (isomorphic to) a complete subalgebra of B_α (this is clear when $\alpha = \beta + 1$, by (2.1)). So (2.2) means, roughly, that the algebra B_α is the direct limit of B_β, $\beta < \alpha$. The exact meaning of (2.2) will be given by the definition of iteration below.

The definition of iteration of length α is by induction on α. For each α, $\dot{Q}_\alpha$ is assumed to be some forcing notion in V^{P_α}, and is assumed to have a greatest element 1. The symbol $\Vdash_\alpha$ denotes the forcing relation corresponding to P_α, and

$$\Vdash_\alpha \sigma$$

means that every condition in P_α forces σ. Similarly, $\leqslant_\alpha$ denotes the partial ordering of P_α.

2.3 Definition

Let $\alpha \geqslant 1$. A forcing notion P_α is an *iteration* (of *length* α, with *finite support*) if it is a set of α-sequences with the following properties:

(i) If $\alpha = 1$ then for some forcing notion Q_0, P_1 is the set of all 1-sequences $\langle p(0) \rangle$, where $p(0) \in Q_0$. And $\langle p(0) \rangle \leqslant_1 \langle q(0) \rangle$ iff $p(0) \leqslant q(0)$ in Q.

(ii) If $\alpha = \beta + 1$ then $P_\beta = \{p \restriction \beta : p \in P_\alpha\}$ is an iteration of length β, and there is some forcing notion $\dot{Q}_\beta \in V^{P_\beta}$ such that

$$p \in P_\alpha \quad \text{iff} \quad p \restriction \beta \in P_\beta \quad \text{and} \quad \Vdash_\beta p(\beta) \in \dot{Q}_\beta$$
$$p \leqslant_\alpha q \quad \text{iff} \quad p \restriction \beta \leqslant_\beta q \restriction \beta \quad \text{and} \quad p \restriction \beta \Vdash_\beta p(\beta) \leqslant q(\beta)$$

(iii) If α is a limit ordinal, then $\forall\beta<\alpha$ $P_\beta = P_\alpha\restriction\beta = \{p\restriction\beta : p\in P_\alpha\}$ is an iteration (of length β), and

$$p\in P_\alpha \text{ iff } \forall\beta<\alpha \quad p\restriction\beta\in P_\beta \text{ and for all but finitely many } \beta<\alpha,$$
$$\Vdash_{\beta} p(\beta)=1$$
$$p\leqslant_\alpha q \text{ iff } \forall\beta<\alpha\, p\restriction\beta\leqslant_\beta q\restriction\beta$$

The finite set $\{\beta<\alpha\colon \text{not} \Vdash_{\beta} p(\beta)=1\}$ is the *support* of a condition $p\in P_\alpha$.

A finite support iteration P_α is uniquely determined by the sequence $\langle\dot{Q}_\beta:\beta<\alpha\rangle$, where each $\dot{Q}_\beta$ is a forcing notion in V^{P_β}, P_β being the initial segment $P_\alpha\restriction\beta$ of the iteration. For each $\beta<\alpha$, $P_{\beta+1}$ is (isomorphic to) the two step iteration $P_\beta*\dot{Q}_\beta$. Thus, we are justified in calling P_α (somewhat imprecisely) the (finite support) *iteration of* $\langle\dot{Q}_\beta:\beta<\alpha\rangle$.

2.4 Lemma

If P_α is a finite support iteration and $P_\beta = P_\alpha\restriction\beta$, then $V^{P_\beta}\subseteq V^{P_\alpha}$.

Proof This can be proved either directly, by showing that if G is generic on P_α then $G_\beta=\{p\restriction\beta : p\in G\}$ is generic on P, or by invoking Lemma 2.7 in Part I. The *canonical projection* $h=h_{\alpha\beta}:P_\alpha\to P_\beta$ and the *canonical embedding* $i=i_{\beta\alpha}:P_\beta\to P_\alpha$ are defined by

$$h(p)=p\restriction\beta, \quad i(q)=q^\frown 1^\frown 1^\frown\ldots \quad \square \tag{2.5}$$

Note that if α is a limit ordinal, then

$$P_\alpha=\bigcup_{\beta<\alpha} i_{\beta\alpha}(P_\beta), \quad \leqslant_\alpha=\bigcup_{\beta<\alpha} i_{\beta\alpha}(\leqslant_\beta) \tag{2.6}$$

which justifies the notation (2.2).

The most important feature of finite support iteration is preservation of chain conditions.

2.7 Theorem

Let κ be a regular uncountable cardinal. Let P_α be a finite support iteration of $\langle\dot{Q}_\beta:\beta<\alpha\rangle$, and let us assume that for each $\beta<\alpha$, $\Vdash_{\beta}\dot{Q}_\beta$ has the κ-c.c. Then P_α has the κ-chain condition.

Proof By induction on α. If $\alpha=\beta+1$ then $P_\alpha=P_\beta*\dot{Q}_\beta$ and the theorem follows from Theorem 1.10. Thus, let α be a limit ordinal. For each $p\in P_\alpha$, let supp(p) denote the support of p.

Case I cf $\alpha \neq \kappa$. Let $W \subseteq P_\alpha$ be a set of size κ. Since cf $\alpha \neq \kappa$, there exists a $\beta < \alpha$ and some $Z \subset W$ of size κ such that $\forall p \in Z$ supp$(p) \subset \beta$. Then $\{p \restriction \beta : p \in Z\}$ is a set of size κ in P_β, and since P_β has the κ-c.c., there are p and q in Z such that $p \restriction \beta$ and $q \restriction \beta$ are compatible (in P_β). But then $p|q$.

Case II cf $\alpha = \kappa$. Let $\{\alpha_\xi : \xi < \kappa\}$ be a normal sequence with limit α, and let $W = \{p_\xi : \xi < \kappa\}$ be a subset of P_α of size κ. For each limit $\xi < \kappa$ there is $\gamma(\xi) < \xi$ such that supp$(p) \cap \alpha_\xi \subset \alpha_{\gamma(\xi)}$. By Fodor's theorem there is a stationary $S \subset \kappa$ and some $\gamma < \kappa$ such that supp$(p_\xi) \cap \alpha_\xi \subset \alpha_\gamma$ for all $\xi \in S$. Furthermore, we easily construct a set $Z \subset S$ of size κ so that for any $\xi < \eta$ in Z, supp$(p_\xi) \subseteq \alpha_\eta$.

Consider the set $\{p_\xi \restriction \alpha_\gamma : \xi \in Z\}$. This is a subset of P_{α_γ}, of size κ, and so there are $\xi < \eta$ in Z such that $p_\xi \restriction \alpha_\gamma$ and $p_\eta \restriction \alpha_\gamma$ are compatible. Let $q \in P_{\alpha_\gamma}$ be stronger than both. Now consider this condition r in P_α:

$$r(i) = \begin{cases} q(i) & \text{if } i < \alpha_\gamma \\ p_\xi(i) & \text{if } \alpha_\gamma \leqslant i < \alpha_\eta \\ p_\eta(i) & \text{if } \alpha_\eta \leqslant i < \alpha \end{cases}$$

The condition r is (in P_α) stronger than both p_ξ and p_η, and so p_ξ and p_η are compatible. □

References

Solovay, R. and Tennenbaum, S. (1971). *Ann. Math.* **94**, 201–45.

Baumgartner, J. (1983). In *Surveys in Set Theory* (Mathias, A.R.D., ed.), pp. 1–59, Cambridge University Press

3 *Martin's Axiom*

The solution of Suslin's problem (see Chapter 4) in the late 1960s, by iterated forcing, led to a formulation of an 'internal forcing axiom' called *Martin's Axiom.* The use of Martin's Axiom often eliminates a separate forcing construction when one tries to establish the consistency of some conjecture. Martin's Axiom has proved to be very useful in combinatorial set theory, and has become particularly popular among point set topologists.

3.1 Martin's Axiom (MA)

For every c.c.c. partial order $(P, <)$, if $\lambda < 2^{\aleph_0}$ and $\{D_\alpha : \alpha < \lambda\}$ is a collection of dense sets in P, then there exists a filter G on P such that $G \cap D_\alpha$ is nonempty, for every $\alpha < \lambda$.

We first remark that MA is a consequence of the continuum hypothesis: if P is any partial order (not necessarily c.c.c.) and $\{D_n : n \in \omega\}$ are dense sets, then it is easy to find a filter that meets all the D_n; let $p_0 \geqslant p_1 \geqslant \cdots \geqslant p_n \geqslant \cdots$ be such that $p_n \in D_n$, and let G be generated by $\{p_n\}_n$.

We shall show that MA is consistent with the negation of the continuum hypothesis.

The often used consequence of MA + $2^{\aleph_0} > \aleph_1$ is Martin's Axiom for collections of $\aleph_1$ dense sets:

3.2 (MA$\aleph_1$)

If P is c.c.c. and if $\{D_\alpha : \alpha < \omega_1\}$ are dense sets, then there exists a filter G on P that meets all the D_α.

3.3

Note that MA$\aleph_1$ implies that $2^{\aleph_0} > \aleph_1$: let P be the Cohen forcing $(P = \{0,1\}^{<\omega})$, and let $\{f_\alpha : \alpha < \omega_1\} \subset \{0,1\}^\omega$. For each α, let $D_\alpha = \{p : p \not\subset f_\alpha\}$. If G meets all the dense sets D_α, then $f = \bigcup G$ differs from all the f_α.

3.4 Lemma

MA is equivalent to the following variant:

MA* If P is c.c.c. and $|P| < 2^{\aleph_0}$, and if $\{D_\alpha : \alpha < \lambda\}, \lambda < 2^{\aleph_0}$ are dense, then there is a filter G that meets all the D_α.

Proof Let us assume MA* and prove MA. Let P be a c.c.c. partial order, and let D_α, $\alpha<\lambda$ ($\lambda<2^{\aleph_0}$) be dense sets in P. It is not difficult to find $Q\subset P$ of size λ such that (i) any $p,q\in Q$ that are compatible in P are compatible in Q, and (ii) for all α, $D_\alpha\cap Q$ is dense in Q. Q is c.c.c. and so we apply MA* to Q and the sets $D_\alpha\cap Q$. The filter we obtain generates a filter on P that verifies MA. □

3.5 Theorem

Assume GCH and let κ be a regular cardinal greater than $\aleph_1$. There is a cardinal preserving notion of forcing P such that the generic extension by P satisfies Martin's Axiom and $2^{\aleph_0}=\kappa$.

Proof We construct P as a finite support iteration of length κ of certain $\{\dot{Q}_\alpha:\alpha<\kappa\}$. At each stage, $\Vdash_{P_\alpha}\dot{Q}_\alpha$ is c.c.c., so P is c.c.c., and so all cardinals are preserved.

The construction is such that $|P_\alpha|\leqslant\kappa$ for each $\alpha\leqslant\kappa$. This property is certainly true for a limit α if it is true for all $\beta<\alpha$, because P_α is the direct limit of the P_β's. At a successor stage, $P_{\alpha+1}=P_\alpha*\dot{Q}_\alpha$, and the $\dot{Q}_\alpha$ will be such that $\Vdash_{\bar{\alpha}}|\dot{Q}_\alpha|<\kappa$. Because P_α has the c.c.c., there is $\lambda<\kappa$ such that $\Vdash_{\bar{\alpha}}|\dot{Q}_\alpha|\leqslant\lambda$. Every name $\dot{q}$ for an element of $\dot{Q}_\alpha$ can be represented by a function from an antichain in P_α into λ; the number of those is $\leqslant\kappa^{\aleph_0}\cdot\lambda^{\aleph_0}=\kappa$ (by c.c.c. again). It follows that $|P_{\alpha+1}|\leqslant\kappa$.

Since GCH holds in the ground model, and because each P_α is a c.c.c. forcing of size $\leqslant\kappa$, we have $\Vdash_{\bar{\alpha}}(\forall\lambda<\kappa)2^\lambda\leqslant\kappa$, for every $\alpha\leqslant\kappa$.

We shall now define the $\dot{Q}_\alpha$, by induction on $\alpha<\kappa$. Let us fix a function π mapping κ onto $\kappa\times\kappa$ so that if $\pi(\alpha)=(\beta,\gamma)$, then $\beta\leqslant\alpha$. For $\alpha<\kappa$, the model V^{P_α} has at most κ distinct (nonisomorphic) partial orderings of size $<\kappa$; as P_α is c.c.c., there are at most κ distinct names in V^{P_α} for such partial orderings. Thus let us assume that $\{\dot{Q}_\beta:\beta<\alpha\}$ have been defined. Let $\pi(\alpha)=(\beta,\gamma)$. Let $\dot{Q}$ be the γth name for a partial order of size κ in the model V^{P_β}. In V^{P_α} we define:

$$\dot{Q}_\alpha=\begin{cases}\dot{Q} & \text{if } \dot{Q} \text{ satisfies c.c.c.}\\ \{1\} & \text{otherwise}\end{cases}$$

The $\dot{Q}_\alpha$, $\alpha<\kappa$, determine the iteration $P=P_\kappa$. Now we have to prove that V^P satisfies MA as well as $2^{\aleph_0}=\kappa$. Let G be a generic filter on P, and let $G_\alpha=G\restriction P_\alpha$ for all $\alpha<\kappa$.

3.6 Lemma

If $\lambda < \kappa$ and $X \subseteq \lambda$ is in $V[G]$, then $X \in V[G_\alpha]$ for some $\alpha < \kappa$.

Proof By the c.c.c., each Boolean value $b_\xi = \| \xi \in \dot{X} \|$ is determined by a countable set of conditions in P (a maximal partition of b_ξ in P); hence, $\dot{X}$ is determined by at most λ conditions in P. Each condition has a finite support and so in fact belongs to some P_α, $\alpha < \kappa$. It follows that there is $\alpha < \kappa$ such that X has a name in V^{P_α}. □

We now prove that MA holds in $V[G]$. By Lemma 3.4 it suffices to prove MA*. Let Q be a c.c.c. partial order in $V[G]$ of size $< \kappa$, and let $\mathscr{D}$ be a collection of dense sets in Q, of size $< \kappa$. As a consequence of Lemma 3.6, both Q and $\mathscr{D}$ are in $V[G_\beta]$ for some $\beta < \kappa$. Let $\dot{Q}$ be a name for Q in V^{P_β}, and let it be the γth name. Let α be such that $\pi(\alpha) = (\beta, \gamma)$. As Q has the c.c.c. in $V[G]$, it has the c.c.c. in $V[G_\alpha]$. Thus, $Q = \dot{Q}_\alpha / G_\alpha$.

In $V[G_{\alpha+1}]$ there is a generic filter H on Q over $V[G_\alpha]$ (because $P_{\alpha+1} = P_\alpha * \dot{Q}_\alpha$). The filter H meets every dense D in Q that is in $V[G_\alpha]$, in particular every $D \in \mathscr{D}$.

In particular, if Q is the Cohen forcing ($\{0,1\}^{<\omega}$), the argument from 3.3 proves that $2^{\aleph_0} \geqslant \kappa$. We have already proved that $2^{\aleph_0} \leqslant \kappa$ and so $V[G]$ satisfies $2^{\aleph_0} = \kappa$. This also means that we have found a G for any c.c.c. Q and $\mathscr{D}$ of size $< 2^{\aleph_0}$, i.e. we proved MA*. Therefore MA holds in $V[G]$. □

Among the many consequences of Martin's Axiom let me mention just two:

3.7 Theorem

Assume MA. Then

(*a*) the union of less than $2^{\aleph_0}$ meagre sets is meagre;

(*b*) the union of less than $2^{\aleph_0}$ Lebesgue null sets in null. □

(Consequently, $2^{\aleph_0}$ is a regular cardinal.)

References

Martin, D.A. and Solovay, R. (1970). *Ann. Math. Logic* **2**, 143–78.

Shoenfield, J. (1975). *Am. Math. Monthly* **82**, 610–17.

Rudin, M.E. (1977). In *Handbook of Mathematical Logic* (Barwise, J., ed.), pp. 491–501, North-Holland, Amsterdam.

Fremlin, D.H. (1984). *Consequences of Martin's Axiom*, Cambridge Tracts in Mathematics 84, Cambridge University Press.

4 *Suslin's problem*

The problem dates back to 1920. Let $(L, <)$ be a dense linear ordering without endpoints, Dedekind complete, and such that it has no uncountable family of disjoint open intervals. Is L isomorphic to the real line?

The eventual solution of the problem led, among other things, to the introduction of iterated forcing and Martin's Axiom. Suslin's problem is independent of the axioms of set theory. A *Suslin line* is a linear ordering that satisfies the assumptions in the problem but is not isomorphic to the real line. There are models of set theory in which a Suslin line exists; in this chapter we show that the existence of a Suslin line is not provable in ZFC. We prove below that $MA\aleph_1$ implies that Suslin lines do not exist.

4.1 Definition

A *Suslin tree* is an uncountable partially ordered set $(T, <)$ such that

(i) for every $t \in T$, the set of all predecessors $\{s \in T : s < t\}$ is well ordered (and we call its order type the *height* of t);

(ii) T has no uncountable chains (linearly ordered subsets) and no uncountable antichains (pairwise incomparable).

Note that the height of each $t \in T$ is a countable ordinal.

4.2 Lemma

A Suslin tree exists if and only if a Suslin line exists.

Proof (*a*) Let $(L, <)$ be a Suslin line. By induction on $\alpha \in \omega_1$, we construct a sequence $\{I_\alpha\}_\alpha$ of open intervals of L as follows: at stage α, choose I_α so that for every $\beta < \alpha$, either I_α is disjoint from I_β, or $I_\alpha \subset I_\beta$. Since L is not isomorphic to the reals, the endpoints of the intervals $\{I_\beta : \beta < \alpha\}$ are not a dense set in L, and so I_α exists.

The set $\{I_\alpha : \alpha < \omega_1\}$, ordered by inverse inclusion $\supseteq$, is a Suslin tree: (i) is clear from the construction, and it is clear as well that T has no uncountable antichain (pairwise disjoint I_α's). T has no uncountable chain either. Suppose that $C \subset \{I_\alpha\}_\alpha$ is an uncountable chain. Then either the left endpoints of the I_α's in C form an increasing sequence $\{a_\eta:\eta<\omega_1\}$ in L of order type ω_1, or the right endpoints form a decreasing sequence of length ω_1. But that is impossible because $\{(a_\eta, a_{\eta+1}):\eta<\omega_1\}$ are disjoint open intervals, and similarly for the right endpoints.

(*b*) Let $(T_1, <)$ be a Suslin tree. Let

$$T=\{t\in T_1:\text{the set }\{s\in T_1:t<s\}\text{ is uncountable}\}$$

T is also a Suslin tree and has the additional property:

$$\text{for every } t\in T, \{s\in t:t<s\} \text{ is uncountable} \tag{4.3}$$

(To prove (4.3), use the fact that for every $\alpha<\omega_1$ there are only countably many $t\in T$ of height α.)

Let L be the set of all *branches* in T, i.e. of all maximal chains. We endow L with a linear order $<$ as follows: for each $\alpha<\omega_1$, we linearly order the set of all $t\in T$ of height α. If b and c are branches in T then let α be the least ordinal such that $b(\alpha)\neq c(\alpha)$, where $b(\alpha)$ and $c(\alpha)$ are the unique $s\in b$ and $t\in c$ of height α (if they exist); such α exists and we let $b<c$ just in case $b(\alpha)<c(\alpha)$.

It is not difficult to see that the linearly ordered set $(L, <)$ is not separable (does not have a countable dense subset). L itself need not be dense, but it has a nonseparable subset that is dense, and its Dedekind completion is a Suslin line. □

Using Suslin trees, the nonexistence of Suslin lines can now be proved by an application of Martin's Axiom:

4.4 Theorem

If MA holds and $2^{\aleph_0}>\aleph_1$, then there are no Suslin trees.

Proof Assume MA$\aleph_1$, and let $(T, <)$ be a Suslin tree. Without loss of generality we may assume that T also satisfies (4.3). Let $(P, <)$ be the partially ordered set obtained by considering T upside down; i.e. T with the reverse order $>$.

We apply MA$\aleph_1$ to the notion of forcing P. If t and s are comparable in T then $t|s$ as forcing conditions; if t and s are *incomparable* in T, then t and s are *incompatible* as forcing conditions. As T is a Suslin tree, P has the countable chain condition.

For each $\alpha<\omega_1$, let

$$D_\alpha=\{t\in T:\text{height }(t)\geqslant\alpha\}$$

By (4.3), each D_α is dense in P. Using MA$\aleph_1$, there is a filter G on P that meets every D_α. It follows that G is an uncountable chain in T, which is a contradiction. □

References

Souslin, M. (1920). *Fund. Math.* **1**, 223.
Solovay, R. and Tennenbaum, S. (1971). *Ann. Math.* **94**, 201–45.
Rudin, M.E. (1969), *Am. Math. Monthly* **76**, 1113–19.
Jech, T. (1971). *J. Symb. Logic* **36**, 1–14.
Devlin, K. and Johnsbråten, H. (1974). *The Souslin Problem.* Lecture Notes in Mathematics 405, Springer-Verlag, NY.

5 Whitehead's problem

In this chapter we describe an application of Martin's Axiom in the theory of infinite abelian groups.

Throughout this chapter, a *group* means an infinite abelian group (with operation $+$). A group is *free* if it has a *basis*, i.e. a set of generators that are linearly independent.

Free groups are characterized by the following property: let A be a free group, and let $\pi: B \to A$ be a homomorphism of some group B onto A. Let X be a basis of A, and for each $x \in X$, pick $\varphi(x) \in B$ so that $\pi(\varphi(x)) = x$. Since X is a basis, φ can be extended to an (injective) homomorphism φ of A into B such that $\pi(\varphi(a)) = a$ for all $a \in A$.

Now let A be an arbitrary group.

5.1 Definition

A is a *W-group* if for any homomorphism $\pi: B \to A$ onto A with kernel $\mathbf{Z}$, there exists a homomorphism $\varphi: A \to B$ such that $\pi(\varphi(a)) = a$ for all $a \in A$.

By the remark preceding Definition 5.1, every free group is a W-group.

5.2 Whitehead's problem

Is every W-group free?

It had been known that every countable W-group is free. Shelah proved that Whitehead's problem is undecidable in set theory. Here we are concerned with the half of his solution that shows that it is consistent that an uncountable W-group exists that is not free:

5.3 Theorem (Shelah)

$MA\aleph_1$ implies that there is a W-group of cardinality $\aleph_1$ that is not free.

We shall outline the idea of Shelah's proof, but first we need some more definitions and facts on W-groups.

5.4 Definition

A group A is *torsion free* if every nonzero $a \in A$ has order ∞, i.e.

$n \cdot a \neq 0$ for all $n = 0, 1, 2, \ldots$. A is $\aleph_1$*-free* if every countable subgroup of A is free.

We note that $\aleph_1$-free implies torsion free because A is torsion free iff every finitely generated subgroup of A is free (this is a theorem in abelian group theory).

It is known that every W-group is $\aleph_1$-free. It had also been known that there exists a group A such that

(i) $|A| = \aleph_1$
(ii) A is not free
(iii) A is $\aleph_1$-free (5.5)
(iv) every countable subgroup of A is included in a countable subgroup B of A such that A/B is $\aleph_1$-free

The proof of Theorem 5.3 consists of showing that under MA$\aleph_1$, the group A in (5.5) is a W-group, and so is a counterexample to Whitehead's problem.

Let A be an $\aleph_1$-free group of size $\aleph_1$ with property (5.5)(iv). A subgroup $S \subset A$ is a *pure subgroup* if A/S is torsion free.

Let π be a homomorphism of some B onto A. We use MA$\aleph_1$ to construct a homomorphism $\varphi: A \to B$ such that $\pi(\varphi(a)) = a$ for all $a \in A$.

Let P be the set of all homomorphisms $\varphi: S \to B$ such that S is a finitely generated pure subgroup of A, and $\pi(\varphi(a)) = a$ for all $a \in S$. We consider P as a notion of forcing, partially ordered by $\supseteq$.

For each $a \in A$, let

$$D_a = \{\varphi \in P : a \in \operatorname{dom}(\varphi)\}$$

We need two things to apply MA$\aleph_1$:

Every D_a is dense in P (5.6)
P satisfies the countable chain condition (5.7)

If (5.6) and (5.7) hold then by MA$\aleph_1$ there is a filter G on P that meets every D_a. Clearly, G yields a homomorphism of A into B verifying that A is a W-group.

It is relatively easy to prove that (5.6) holds. The proof requires a little bit of group theory and only uses the fact that A is $\aleph_1$-free.

The heart of the proof lies in showing that P has the countable chain condition. One employs a Δ-system argument, but the proof uses substantially the property (iv) from (5.5).

References

Shelah, S. (1974), *Israel J. Math.* **18**, 243–56.
Eklof, P. (1976). *Am. Math. Monthly* **83**, 775–88.

6 Kaplansky's conjecture

This chapter deals with an application of forcing in the theory of Banach algebras. $C[0,1]$ is the algebra of continuous functions on $[0,1]$.

Kaplansky's problem

Is every homomorphism on $C[0,1]$ (into any commutative Banach algebra) continuous?

Under the assumption of CH there exist homomorphisms on $C[0,1]$ that are discontinuous. We devote this chapter to the following independence result:

6.1 Theorem (Solovay, Woodin)

It is consistent that every homomorphism on $C[0,1]$ is continuous.

Kaplansky's conjecture had been successively reduced to a purely set theoretic problem. For functions $f, g \in \omega^\omega$, we define (a partial ordering)

$$f < g \quad \text{iff} \quad \exists k \quad \forall n > k \quad f(n) < g(n) \tag{6.2}$$

Similarly, if U is an ultrafilter on ω, we let

$$f <_U g \quad \text{iff} \quad \{n : f(n) < g(n)\} \in U \tag{6.3}$$

For $g \in \omega^\omega$, we denote by

$$\mathrm{Ult}_U g$$

the initial segment $\{f/U : f <_U g\}$ of the linearly ordered set $(\omega^\omega/U; <_U)$. A function $g \in \omega^\omega$ is *monotone* if $m < n$ implies $g(m) \leqslant g(n)$.

6.4 Theorem (Woodin)

If there is a discontinuous homomorphism then there exists a nonprincipal ultrafilter U on ω, an unbounded monotone function g and an order-preserving function

$$\pi : \mathrm{Ult}_U(g) \to (\omega^\omega, <) \qquad \square$$

We shall construct a model in which Woodin's necessary condition fails, and consequently there is no discontinuous homomorphism in that model.

The proof is in two steps, one using MA and the other constructing a model by iterated forcing. Consider the set $\{0,1\}^{\omega_1}$ ordered lexicographically.

6.5 Lemma

Assume MA + not CH. For every nonprincipal U and every monotone unbounded g, $\{0,1\}^{\omega_1}$ can be embedded in $\mathrm{Ult}_U g$.

6.6 Lemma

There is a model of MA + not CH which satisfies that $\{0,1\}^{\omega_1}$ cannot be embedded in $(\omega^\omega, <)$.

These two lemmas and Theorem 6.4 establish the consistency result 6.1. We prove Lemma 6.5 first.

6.7 Lemma

Let $\{f_n\}_n$ and $\{g_n\}_n$ be two sequences of functions in ω^ω such that

$$f_0 > f_1 > \cdots > f_n > \cdots > g_n > \cdots > g_1 > g_0$$

Then there are f and g in ω^ω such that for all n,

$$f_n > f > g > g_n$$

and, moreover, $\lim_k(f(k) - g(k)) = \infty$.

Proof We find $k_0 < k_1 < \cdots < k_n < \cdots$ such that for each n, k_n is so that $\forall k \geqslant k_n$,

$$f_0(k) > \cdots > f_n(k) > g_n(k) > \cdots > g_0(k)$$

Let f and g be so that for all n, if $k_n \leqslant k < k_{n+1}$, $f(k) = f_n(k)$ and $g(k) = g_n(k)$. □

6.8 Lemma

Assume MA$\aleph_1$. Let $\{f_\alpha\}_\alpha$ and $\{g_\alpha\}_\alpha$ be two ω_1-sequences of functions in ω^ω such that

$$f_0 > \cdots > f_\alpha > \cdots > g_\alpha > \cdots > g_0$$

For every nonprincipal ultrafilter U on ω there exists $h \in \omega^\omega$ such that for all $\alpha < \omega_1$,

$$f_\alpha >_U h >_U g_\alpha$$

Proof Consider the following notion of forcing P: a condition $p \in P$ is a triple (E, h_1, h_2), where

(6.9) (i) $E=\{\alpha_1,\ldots,\alpha_m\}$ is a finite subset of ω_1, $\alpha_1<\cdots<\alpha_m$
(ii) $h_1=\langle h_1(0),\ldots,h_1(k_p)\rangle$ and $h_2=\langle h_2(0),\ldots,h_2(k_p)\rangle$ are finite functions (into ω), and
(iii) k_p is such that for every $k>k_p$

$$f_{\alpha_1}(k)>\cdots>f_{\alpha_m}(k)>g_{\alpha_m}(k)>\cdots>g_{\alpha_1}(k) \qquad (*)$$

A condition $q=(\bar{E},\bar{h}_1,\bar{h}_2)$ is stronger than $p=(E,h_1,h_2)$ if:

(iv) $\bar{E}\supseteq E$
(v) $\bar{h}_1$ extends h_1, $\bar{h}_2$ extends h_2, and
(vi) $\forall\alpha\in E\forall k>k_p(k\leqslant k_q)$ either $g_\alpha(k)<\bar{h}_1(k)<f_\alpha(k)$ or $g_\alpha(k)<\bar{h}_2(k)<f_\alpha(k)$.

Claim P satisfies the c.c.c.

Proof As there are only countably many pairs (h_1,h_2), it suffices to show that any two conditions with the same h_1 and h_2 are compatible. Let $p=(E,h_1,h_2)$ and $q=(F,h_1,h_2)$. For all $k>k_p=k_p=k_q=\text{length}(h_i)$, the functions $\{f_\alpha,g_\alpha:\alpha\in E\}$ satisfy (6.9)($*$), and so do the functions $\{f_\alpha,g_\alpha:\alpha\in F\}$. Let k_r be such that all the $\{f_\alpha,g_\alpha:\alpha\in E\cup F\}$ have the property ($*$) for all $k>k_r$.

We extend h_1 and h_2 to $\bar{h}_1$ and $\bar{h}_2$ (of length k_r) so that for every k between k_p and k_r, if $\alpha\in E$ then $f_\alpha(k)>\bar{h}_1(k)>g_\alpha(k)$, and if $\alpha\in F$ then $f_\alpha(k)>\bar{h}_2(k)>g_\alpha(k)$. It follows that the condition $r=(E\cup F,\bar{h}_1,\bar{h}_2)$ is stronger than both p and q. □

Now let

$$D_n=\{p\in P:\kappa_p\geqslant n\} \quad (n\in\omega)$$

and

$$\Delta_\alpha=\{p\in P:\alpha\in E \quad \text{where} \quad p=(E,h_1,h_2)\} \quad (\alpha<\omega_1)$$

Each D_n and each Δ_α is dense. By MA$\aleph_1$ there is a filter G on P that meets all the D_n and all the Δ_α. G yields a pair of functions $h_1,h_2\in\omega^\omega$.

If $\alpha<\omega_1$ then there is a $p\in G$ such that $\alpha\in E$, $p=(E,h_1\restriction k_p,h_2\restriction k_p)$. For every $k>k_p$, either $f_\alpha(k)>h_1(k)>g_\alpha(k)$, or $f_\alpha(k)>h_2(k)>g_\alpha(k)$. It follows that either (1) $f_\alpha>_U h_1>_U g_\alpha$, or (2) $f_\alpha>_U h_2>_U g_\alpha$. Either (1) or (2) holds for uncountably many α's, let us say (1), but then (1) holds for all α, proving the lemma. □

Proof of Lemma 6.5

Let U be a nonprincipal ultrafilter on ω, and let $g\in\omega^\omega$ be unbounded and monotone. To each $t\in\{0,1\}^{<\omega_1}$, we assign a pair of functions

f_t and g_t so that $\lim_k(f_t(k) - g_t(k)) = \infty$, as follows (by induction on the length of t): first let $g_\varnothing < f_\varnothing < g$. Given f_t and g_t, we find f_{t0}, g_{t0}, and f_{t1}, g_{t1} so that

$$g_t < g_{t0} < f_{t0} < g_{t1} < f_{t1} < f_t$$

If t has limit length, we apply Lemma 6.7 to find f_t and g_t so that for every $s \subset t$, $g_s < g_t < f_t < f_s$.

If $b \in \{0,1\}^{\omega_1}$, then $\{f_t : t \subset b\}$ and $\{g_t : t \subset b\}$ satisfy the assumption of Lemma 6.8. There is therefore some $h_b = h$ such that $f_t >_U h >_U g_t$ for all $t \subset b$.

Now it is easy to verify that if $b_1 < b_2$ in $\{0,1\}^{\omega_1}$, then $h_{b_1} <_U h_{b_2}$ and thus $\{0,1\}^{\omega_1}$ is embedded in $\mathrm{Ult}_U g$. □

Toward proving Lemma 6.6, let

$$F = \{f_\alpha : \alpha < \omega_1\}, \quad G = \{g_\alpha : \alpha < \omega_1\}$$

be such that F is decreasing in $(\omega^\omega, <)$, and G is increasing, and such that $F > G$, i.e. $f_\alpha > g_\alpha$ for all α.

6.10 Definition

(F, G) is a *gap* if there is no $h \in \omega^\omega$ such that $F > h > G$. (F, G) is a *strong gap* if

(i) for every α, $f_\alpha(k) > g_\alpha(k)$ for all $k \in \omega$

(ii) for all $\alpha \neq \beta$, either $f_\alpha(k) \leqslant g_\beta(k)$ for some k, or $f_\beta(k) \leqslant g_\alpha(k)$ for some k.

6.11 Lemma

A strong gap is a gap.

Proof Assume that $F > h > G$. There is an uncountable $W \subset \omega_1$ and some fixed k_0 such that $\forall k \geqslant k_0$, $\forall \alpha \in W$, $f_\alpha(k) > h(k) > g_\alpha(k)$. Furthermore, we may assume that for all $\alpha \in W$, the f_α's have the same initial segment $f_\alpha \restriction k_0$, and similarly the $g_\alpha \restriction k_0$ are all the same.

Now (i) and (ii) lead to a contradiction: let $\alpha \neq \beta$ be in W. By (i) we have $f_\alpha(k) > g_\alpha(k) = g_\beta(k)$ for all $k < k_0$, and for $k \geqslant k_0$ we have $f_\alpha(k) > h(k) > g_\beta(k)$. Similarly, $f_\beta(k) > g_\alpha(k)$ for all k; a contradiction. □

Being a gap is of course not an absolute property for models of set theory, since (F, G) may be a gap in a smaller model, but not be a gap in a larger model. Being a strong gap, however, is an absolute property, as long as $\aleph_1$ remains $\aleph_1$. This fact is used in the proof of Lemma 6.6 below.

6.12 Lemma

Let (F,G) be a gap. There is a c.c.c. notion of forcing such that in the generic extension there is a strong gap $(\bar{F},\bar{G})$ such that for every $\alpha<\omega_1$ there is k_0 so that

$$\bar{f}_\alpha(k)=f_\alpha(k) \quad \text{and} \quad \bar{g}_\alpha(k)=g_\alpha(k), \quad \text{for all} \quad k\geqslant k_0 \tag{6.13}$$

Proof A condition is a finite set $E\subset\omega_1$ and functions $\bar{f}_\alpha$, $\bar{g}_\alpha$, $\alpha\in E$, that satisfy (6.13), and also satisfy (i) and (ii) from the definition of strong gap. As long as P satisfies the c.c.c., $\aleph_1$ is preserved and so a generic filter on P clearly yields a strong gap that satisfies (6.13).

In order to show that P satisfies the c.c.c., let $W\subset P$ be uncountable. W contains an uncountable Δ-system, with root $A\subset\omega_1$; i.e. conditions p_ξ, $\xi<\omega_1$, so that if $\xi<\eta$ then $E_\xi\cap E_\eta=A$, and $\max E_\xi<\min(E_\xi-A)$. We shall find ξ and η so that p_ξ and p_η are compatible.

For each ξ, let

$$f^*_\xi=\max\{\bar{f}_\alpha:\alpha\in E_\xi-A\}, \quad g^*_\xi=\min\{\bar{g}_\alpha:\alpha\in E_\xi-A\}$$

Claim There are $\xi\neq\eta$ so that $f^*_\xi(k)\leqslant g^*_\eta(k)$ for some k.

Proof Otherwise $\forall\xi,\eta\ \forall k\ f^*_\xi(k)>g^*_\eta(k)$, and then it is easy to see that the function h defined by $h(k)=\sup\{g^*_\xi(k):\xi<\omega_1\}$ exists and is such that $F>h>G$; a contradiction. □

Now if ξ,η are as in the Claim, p_ξ and p_η are compatible: the condition $\{\bar{f}_\alpha,\bar{g}_\alpha:\alpha\in E_\xi\cup E_\eta\}$ satisfies (6.10)(i) and (ii) and is stronger than both p_ξ and p_η. □

Proof of Lemma 6.6

Assume GCH in the ground model. We construct a generic extension in which MA$\aleph_1$ holds, and in which $\{0,1\}^{\omega_1}$ can be embedded in $(\omega^\omega,<)$. We construct the model by a finite support iteration P of length $\aleph_2$. The proof follows closely the construction of a model of Martin's Axiom in Theorem 3.5. At each stage α of the iteration, we consider some c.c.c. forcing notion $\dot{Q}_\alpha$ of size $\aleph_1$. The resulting iteration satisfies the c.c.c., and the generic model satisfies $2^{\aleph_0}\leqslant\aleph_2$.

At even stages of the iteration, we consider all potential c.c.c. notions of forcing, as we did in the proof of Theorem 3.5. At odd stages, we choose $\dot{Q}_\alpha$ in such a way that the final model will not have an embedding of $(\{0,1\}^{\omega_1})^V$ into ω^ω.

Let D be the set of all eventually constant sequences $a\in\{0,1\}^{\omega_1}$. D is dense in $\{0,1\}^{\omega_1}$ and has size $\aleph_1$. There are $\aleph_2$ names $\dot{F} \in V^P$ for an order-preserving map of D into ω^ω in V^P; let $\dot{F}_\alpha$, $\alpha < \omega_2$, be some enumeration so that $\dot{F}_\alpha \in V^{P_\alpha}$ for all α.

Let $\alpha = \beta \cdot 2 + 1$. In V^{P_α}, we define $\dot{Q}_\alpha$ as follows: let $\pi = \dot{F}_\beta$; π is an order-preserving map of D into ω^ω. For each $b \in (\{0,1\}^{\omega_1})^V - D$, let $\{d_\xi\}_\xi$ and $\{e_\xi\}_\xi$ be an increasing and a decreasing ω_1-sequence in D with limit b. Let $F = \pi(\{e_\xi\}_\xi)$ and $G = \pi(\{d_\xi\}_\xi)$.

Claim There is a b such that (F, G) is a gap.

Proof Otherwise, for every b there is $h_b = h \in \omega^\omega$ such that $F > h > G$. If $b_1 < b_2$ then $b_1 < d < b_2$ for some $d \in D$, and so $h_{b_1} < \pi(d) < h_{b_2}$. But $\{0,1\}^{\omega_1}$ has size $\aleph_2$ while ω^ω (in V^{P_α}) has size $\aleph_1$; a contradiction. □

We let $\dot{Q}_\alpha$ be the forcing notion described in Lemma 6.12. It follows that in $V^{P_{\alpha+1}}$ there is a strong gap $(\bar{F}, \bar{G})$ such that (6.13) holds. Consequently, $(\bar{F}, \bar{G})$ is a strong gap in V^P, and so in V^P there can be no $h \in \omega^\omega$ such that $F > h > G$. It follows that $\pi = \dot{F}_\beta$ cannot be extended in V^P to an embedding of $\{0,1\}^{\omega_1}$ into ω^ω.

Therefore no $\dot{F}_\beta$ can be extended in V^P to an embedding of $\{0,1\}^{\omega_1}$ into ω^ω, and Lemma 6.6 is proved. □

References

Dales, H.G. (1979). *Am. J. Math.* **101**, 647–734.
Esterle, J. (1978). *Proc. London Math. Soc.* **36**, 46–58.
Woodin, W.H. (1987). *Memoirs Am. Math. Soc.* (in press).
Dales, H.G. and Woodin, W.H. (1986). *An Introduction to Independence for Analysts*, London Mathematical Society Lecture Notes 115, Cambridge University Press.

7 *Countable support iteration*

Iteration of forcing with finite support was introduced in Chapter 2. We shall now give a general definition of iteration. We closely follow the notation of Chapter 2.

7.1 Definition

Let $\alpha \geqslant 1$. A forcing notion P_α is an *iteration* (of *length* α) if it is a set of α-sequences with the following properties:

(i) If $\alpha = 1$ then for some forcing notion Q_0, P_1 is the set of all 1-sequences $\langle p(0)\rangle$, where $p(0) \in Q_0$. And $\langle p(0)\rangle \leqslant_1 \langle q(0)\rangle$ iff $p(0) \leqslant q(0)$ in Q_0.

(ii) If $\alpha = \beta + 1$ then $P_\beta = P_\alpha \restriction \beta = \{p \restriction \beta : p \in P_\alpha\}$ is an iteration of length β, and there is some forcing notion $\dot{Q}_\beta \in V^{P_\beta}$ such that

$$p \in P_\alpha \quad \text{iff} \quad p \restriction \beta \in P_\beta \quad \text{and} \quad \Vdash_{\beta} p(\beta) \in \dot{Q}_\beta$$
$$p \leqslant_\alpha q \quad \text{iff} \quad p \restriction \beta \leqslant_\beta q \restriction \beta \quad \text{and} \quad p \restriction \beta \Vdash_{\beta} p(\beta) \leqslant q(\beta)$$

(iii) If α is a limit ordinal, then $\forall \beta < \alpha$, $P_\beta = P_\alpha \restriction \beta = \{p \restriction \beta : p \in P_\alpha\}$ is an iteration (of length β), and

(*a*) $1 \in P_\alpha$ (where $1(\xi) = 1$ for all $\xi < \alpha$)

(*b*) if $p \in P_\alpha$, $\beta < \alpha$ and $q \in P_\beta$ is such that $q \leqslant_\beta p \restriction \beta$, then $r \in P_\alpha$, where

$$r(\xi) = \begin{cases} q(\xi) & \text{for} \quad \xi < \beta \\ p(\xi) & \text{for} \quad \beta \leqslant \xi < \alpha. \end{cases}$$

And

$$p \leqslant_\alpha q \quad \text{iff} \quad \forall \beta < \alpha \quad p \restriction \beta \leqslant_\beta q \restriction \beta$$

7.2 Lemma

If P_α is an iteration and $P_\beta = P_\alpha \restriction \beta$, then $V^{P_\beta} \subseteq V^{P_\alpha}$.

Proof As in Lemma 2.4 (using (iii)(*a*) and (*b*)). □

A finite support iteration is a special case of iteration of forcing. An iteration depends in general not only on the forcing notions $\dot{Q}_\beta$, but also on what happens at the limit stages of the iteration.

Let α be a limit ordinal, and let P_α be an iteration of length α. P_α is a *direct limit* if, for every α-sequence p,

$$p \in P_\alpha \quad \leftrightarrow \quad \exists \beta < \alpha \quad (p \upharpoonright \beta \in P_\beta \text{ and } \forall \xi \geqslant \beta \ p(\xi) = 1) \tag{7.3}$$

P_α is an *inverse limit* if, for every α-sequence p,

$$p \in P_\alpha \quad \leftrightarrow \quad \forall \beta < \alpha \quad p \upharpoonright \beta \in P_\beta \tag{7.4}$$

Most common iterations of forcing combine direct are inverse limits. Finite support iterations are exactly those that use only direct limits at all limit ordinals.

If $p \in P_\alpha$, let

$$\operatorname{supp}(p) = \{\beta < \alpha : \text{not } \Vdash_\beta p(\beta) = 1\} \tag{7.5}$$

be the *support* of p. If I is an ideal on α, then an *iteration with I-support* is an iteration P_α that for each limit $\gamma \leqslant \alpha$ satisfies that for every γ-sequence p,

$$p \in P_\gamma \quad \leftrightarrow \quad \forall \beta < \gamma \quad p \upharpoonright \beta \in P_\beta \quad \text{and} \quad \operatorname{supp}(p) \in I$$

Of particular interest are countable support iterations.

7.6 Definition

P_α is a *countable support iteration* if for all limit ordinals $\gamma \leqslant \alpha$,

$$p \in P_\gamma \quad \text{iff} \quad \forall \beta < \alpha \quad p \upharpoonright \beta \in P_\beta \text{ and for all but countably many } \beta < \gamma, \Vdash_\beta p(\beta) = 1$$

Equivalently, it is an iteration such that for all limit $\gamma \leqslant \alpha$,

(*a*) if $\operatorname{cf} \gamma = \omega$ then P_γ is an inverse limit, and
(*b*) if $\operatorname{cf} \gamma > \omega$ then P_γ is a direct limit.

A countable support iteration is determined by the sequence $\{\dot{Q}_\beta\}_{\beta < \alpha}$; in fact, it is the Boolean algebras $B(\dot{Q}_\beta)$ that determine an iteration:

7.7 Lemma

Let P and P' be countable support iterations of length α, of $\{\dot{Q}_\beta\}_\beta$ and $\{\dot{Q}'_\beta\}_\beta$, respectively. Assume that for every $\beta < \alpha$, if $B(P_\beta) = B(P'_\beta)$ then $\Vdash_\beta B(\dot{Q}_\beta) = B(\dot{Q}'_\beta)$. Then $B(P) = B(P')$.

Proof By induction on α; assume that $B(P_\beta) = B(P'_\beta)$ for all $\beta < \alpha$. If α is a successor $\alpha = \beta + 1$ then $B(P_\alpha) = B(P_\beta * \dot{Q}_\beta) = B(P'_\beta * \dot{Q}'_\beta) = B(P'_\alpha)$. So let α be a limit ordinal. Without loss of generality we assume that $\dot{Q}'_\beta = \dot{B}_\beta = (B(\dot{Q}_\beta))^{V^{P_\beta}}$, for all $\beta < \alpha$.

We show that P_α is dense in P'_α. Let $p' \in P'_\alpha$. We construct $p \in P_\alpha$ so that for all $\beta < \alpha$, $p \restriction \beta \leqslant_\beta p' \restriction \beta$. Let $\beta < \alpha$ and assume that we have constructed $p \restriction \beta \leqslant p' \restriction \beta$. If $p' \restriction \beta \models p'(\beta) = 1$, let $p(\beta) = 1$. Otherwise, $p \restriction \beta \models_\beta (\exists q \in \dot{Q}_\beta)$ $q \leqslant p'(\beta)$, hence $\exists \dot{q}$ such that $p \restriction \beta \models_\beta \dot{q} \leqslant p'(\beta)$; let $p(\beta) = \dot{q}$. Clearly, $p \in P_\alpha$ is as desired. □

We now turn our attention to preservation of cardinals under iterations. The following easy lemma shows that a countable support iteration of countably closed notions of forcing is countably closed:

7.8 Lemma

Let κ be a regular cardinal, and let I be an ideal on a limit ordinal α, closed under unions of κ sets. Let P_α be an iteration with I-support. If for each $\beta < \alpha$, $\Vdash_{\overline{\beta}} \dot{Q}_\beta$ is κ-closed, then P_α is κ-closed.

Proof Let $\{p_\xi : \xi < \lambda\}$ be a decreasing sequence in P_α of length $\lambda \leqslant \kappa$. We find a lower bound $p = \langle p(\beta) : \beta < \alpha \rangle$ as follows, by induction on $\beta < \alpha$. Having constructed $p \restriction \beta \in P_\beta$, we let $p(\beta)$ be (in V^{P_β}) a lower bound of $\{p_\xi(\beta) : \xi < \lambda\}$; if all the $p_\xi(\beta)$, $\xi < \lambda$, are $= 1$, then we let $p(\beta) = 1$. This guarantees that $\operatorname{supp}(p) \subseteq \bigcup_\xi \operatorname{supp}(p_\xi)$ and so p is a condition. □

The next theorem deals with the chain condition.

7.9 Theorem

Let κ be a regular uncountable cardinal. Let P_α be an iteration such that each $P_\beta = P_\alpha \restriction \beta$, $\beta < \alpha$, has the κ-chain condition, and that P_α is a direct limit. Furthermore, assume that if $\operatorname{cf} \alpha = \kappa$ then for a stationary set of $\beta < \alpha$, P_β is a direct limit. Then P_α has the κ-chain condition.

Proof The proof is exactly like the proof of Theorem 2.7. Since P_α is a direct limit, $\operatorname{supp}(p)$ is not cofinal in α, for any $p \in P_\alpha$. The proof of Case I ($\operatorname{cf} \alpha \neq \kappa$) holds verbatim; in Case II ($\operatorname{cf} \alpha = \kappa$), we restrict ourselves to a stationary set $S_0 \subset \kappa$ such that for every $\xi < \kappa$, P_{α_ξ} is a direct limit (and so $\operatorname{supp}(p_\xi) \cap \alpha_\xi$ is bounded below α_ξ). □

7.10 Corollary

Let κ be a regular uncountable cardinal, $\kappa \geqslant \aleph_2$. Let P be a countable support iteration of length κ such that for all $\beta < \kappa$, $P \restriction \beta$ has a dense subset of size $< \kappa$. Then P has the κ-chain condition.

Proof P is a direct limit, and so is $P \restriction \beta$ when $\operatorname{cf} \beta > \omega$. But $\{\beta < \kappa : \operatorname{cf} \beta > \omega\}$ is a stationary set. □

In practice, a countable support iteration is used either for $\kappa = \aleph_2$, or so that κ becomes $\aleph_2$ (and cardinals between $\aleph_2$ and κ are collapsed). The main problem is then how to preserve $\aleph_1$.

We shall now prove the *Factor Lemma* for countable support iteration. The lemma states that under certain reasonable assumptions, a countable support iteration $P_{\alpha+\eta}$ (of $\{\dot{Q}_\beta\}_\beta$) is equivalent to $P_\beta * \dot{P}_\eta^{(\alpha)}$, where $\dot{P}_\eta^{(\alpha)}$ is (in V^{P_α}) the countable support iteration of $\{\dot{Q}_\beta\}_{\beta \geq \alpha}$.

This formulation is not quite accurate. The name $\dot{Q}_{\alpha+\beta}$ is an element of $V^{P_{\alpha+\beta}}$, while $\dot{P}_\eta^{(\alpha)}$ is claimed to be inside V^{P_α}, an iteration of some $\dot{Q}_\beta^{(\alpha)}$ that are, technically, Boolean-valued names considered in V^{P_α}. However, the factor lemma establishes (for every η) an isomorphism between $V^{P_{\alpha+\eta}}$ and the Boolean-valued model $V^{\dot{P}_\eta^{(\alpha)}}$ defined in V^{P_α}, and so we can identify $\dot{Q}_{\alpha+\beta} \in V^{P_{\alpha+\beta}}$ and the V^{P_α}-name for $\dot{Q}_\beta^{(\alpha)} \in V^{\dot{P}_\beta^{(\alpha)}}$.

We remark that a similar factor lemma holds for many other kinds of iterated forcing.

For $\alpha < \beta$, we let $[\alpha, \beta)$ be the interval $\{\xi : \alpha \leq \xi < \beta\}$. Let

$$P_{\alpha\beta} = P_\beta \restriction [\alpha, \beta) = \{p \restriction [\alpha, \beta) : p \in P_\beta\} \tag{7.11}$$

In V^{P_α}, let $\dot{<}$ be the partial ordering of $P_{\alpha\beta}$ defined as follows: let $\dot{G}$ be the generic ultrafilter on P_α.

$$p \dot{\leq} q \quad \text{iff for some } r \in \dot{G},\ \widehat{rp} \leq \widehat{rq} \tag{7.12}$$

It is easy to see that P_β is a dense subset of $P_\alpha * (P_{\alpha\beta}, \dot{<})$; thus $(P_{\alpha\beta}, \dot{<})$ is essentially $P_\beta : P_\alpha$.

7.13 The Factor Lemma

Let $P_{\alpha+\eta}$ be a countable support iteration of $\{\dot{Q}_\xi\}_{\xi < \alpha+\eta}$. Assume that the model V^{P_α} has the property that every countable set X of ordinals is included in a set $A \in V$ that is countable in V. Then in the model V^{P_α}, $P_{\alpha,\alpha+\eta}$ is isomorphic to $\dot{P}_\eta^{(\alpha)}$, the countable support iteration of $\{\dot{Q}_{\alpha+\beta}\}_{\beta<\eta}$.

Corollary $B(P_{\alpha+\eta}) = B(P_\alpha * \dot{P}_\eta^{(\alpha)})$. In other words, a countable support iteration $P_{\alpha+\eta}$ of length $\alpha+\eta$ is equivalent to a two-step iteration, of a countable support iteration P_α, followed by a countable support iteration $\dot{P}_\eta^{(\alpha)}$ in V^{P_α}.

Proof We work in $V[G]$, where G is a generic filter on P_α. We show, by induction on η, that $P_{\alpha,\alpha+\eta}$ is isomorphic to $P_\eta^{(\alpha)}$. If $\eta = 1$, then $P_{\alpha,\alpha+1}$ is isomorphic to Q_α, which is isomorphic to $P_1^{(\alpha)}$. In general, we assume that for all $\beta < \eta$, $P_{\alpha,\alpha+\beta}$ is isomorphic to $P_\beta^{(\alpha)}$, so $\dot{Q}_{\alpha+\beta} \in (V[G])^{P_\beta^{(\alpha)}}$, and so $P_\eta^{(\alpha)}$ is defined. If η is a successor ordinal, $\eta = \beta + 1$, then it is easy to show (in $V[G]$) that $P_{\alpha,\alpha+\beta+1}$ is isomorphic to $P_{\alpha,\alpha+\beta} * \dot{Q}_\beta$. Thus, let η be a limit ordinal. We find an isomorphism between $P_\eta^{(\alpha)}$ and $P_{\alpha,\alpha+\eta}$ by assigning to each $q \in P_\eta^{(\alpha)}$ the appropriate corresponding element of $P_{\alpha,\alpha+\eta}$. This correspondence is order preserving, and is onto $P_{\alpha,\alpha+\eta}$.

Let $q \in P_\eta^{(\alpha)}$, and let $\dot{q} \in V^{P_\alpha}$ be a name for q. Now we work in V. For each $\beta < \eta$, let $\dot{q}_\beta$ be, with P_α-value 1, the βth term of the sequence $\dot{q}$. As $\dot{q}_\beta$ can be regarded as being such that $\Vdash_{\overline{\alpha+\beta}} \dot{q}_\beta \in \dot{Q}_{\alpha+\beta}$, we let $p_{\alpha+\beta} = \dot{q}_\beta$. This defines the sequence $p = \langle p_{\alpha+\beta} : \beta < \eta \rangle$.

In V^{P_α}, $\dot{q}$ has a countable support. By the assumption of the lemma, there is a countable set $A \subseteq \eta$ and a condition $r \in G$ that forces that $\mathrm{supp}(\dot{q}) \subseteq A$. It follows that $\widehat{r \langle p_{\alpha+\beta} : \beta < \eta \rangle}$ is such that for all $\beta \notin A$, $\widehat{rp} \restriction (\alpha + \beta) \Vdash_{\overline{\alpha+\beta}} p_{\alpha+\beta} = 1$. Thus, we replace each $p_{\alpha+\beta}$ $(\beta \notin A)$ by 1; now $\mathrm{supp}(\widehat{rp}) \subseteq \{\alpha + \beta : \beta \in A\}$ and so $\widehat{rp} \in P_\alpha$.

Back in $V[G]$, p belongs to $P_{\alpha,\alpha+\eta}$, and the correspondence of q and p is an isomorphism between $P_\eta^{(\alpha)}$ and $P_{\alpha,\alpha+\eta}$. □

In the next chapter we present one application of countable support iteration of forcing, the consistency of Borel's conjecture. We mention another application of countable support iteration. We recall that an ultrafilter U on ω is a *p-point* if for every sequence $\{A_n\}_n$ of elements of U there is $A \in U$ so that each $A - A_n$ is finite. A p-point can be easily constructed when one assumes the continuum hypothesis.

7.14 Theorem (Shelah)

There is a generic model in which $2^{\aleph_0} = \aleph_2$ and there are no p-points.

Shelah's proof uses a countable support iteration of length ω_2. The basic idea is that if U is p-point, if $P(U)$ is the corresponding Grigorieff's forcing, and if $Q = Q_U = P(U)^\omega$ is the (full) product of ω copies of $P(U)$, then in V^Q, U cannot be extended to a p-point.

The iteration uses $\{\dot{Q}_\beta\}_{\alpha < \omega_2}$ so that $\Vdash_\beta \dot{Q}_\beta = Q_U$ for some p-point U, and so that all p-points in all $V^{P_\alpha}, \alpha < \omega_2$, are considered. Each stage $\alpha < \omega_2$ of the iteration preserves $\aleph_1$ and has the $\aleph_2$-chain condition, and satisfies the continuum hypothesis (proofs of these facts are highly non-trivial). For every potential p-point U in $V[G]$ there is $\alpha < \omega_2$ so that $U \cap V[G] \in V[G]$ is considered at stage α (using $\diamond$ that is assumed to hold in the ground model).

The crucial property of this iteration is that when $\dot{Q}_\beta = Q_U$ then U cannot be extended to a p-point, not just in $V[G \restriction (\beta + 1)]$, but in $V[G]$.

7.15 Iteration with Easton support

This is a type of iteration useful when dealing with large cardinals. Every condition has support $S = \mathrm{supp}(p)$ with the property

$$|S \cap \gamma| < \gamma \text{ for every regular cardinal } \gamma$$

A typical application is Silver's consistency proof of $2^\kappa > \kappa^+$ for a measurable cardinal κ (using a supercompact cardinal in the ground model).

References

For more on Easton support iteration, and iterations in general see

Menas, T. (1976). *Trans. Am. Math. Soc.* **223**, 61–91.

Baumgartner, J. (1983). In *Surveys in Set Theory* (Mathias, A.R.D., ed.), pp. 1–59. Cambridge University Press.

8 Borel's conjecture

One application of iteration of forcing with countable support is Laver's proof of consistency of Borel's conjecture.

Let X be a subset of $[0,1]$. X has *strong measure zero* if for every sequence $\{\varepsilon_n\}_{n<\omega}$ of positive reals there exists a sequence $\{I_n\}_n$ of intervals with length $(I_n) \leqslant \varepsilon_n$ such that $X \subseteq \bigcup_{n=0}^{\infty} I_n$.

Clearly, every countable X has strong measure zero, and every strong measure zero has Lebesque measure 0.

8.1 Borel's conjecture

All strong measure zero sets are countable.

Borel's conjecture is unprovable in ZFC, since counterexamples exist under the assumption of $2^{\aleph_0} = \aleph_1$. Laver proved its consistency by constructing a generic model in which Borel's conjecture holds and $2^{\aleph_0} = \aleph_2$.

The model is obtained by the countable support iteration of length ω_2 of $\{\dot{Q}_\beta\}_\beta$, where each $\dot{Q}_\beta$ is the Laver forcing (Part I, Section 3.16). Below we give a sketch of the proof. The central idea of the proof is the following.

8.2 Lemma

Let P be the Laver forcing and let G be generic on P. Every set X of reals in the ground model that has strong zero in $V[G]$ is countable.

In fact, we will show this: Let $f \in V[G]$ be the Laver real, and let $\varepsilon_n = 1/f(n)$, for all n. If $X \in V$ is uncountable then, for some n_0, the sequence $\{\varepsilon_n\}_{n \geqslant n_0}$ is a counterexample to X having strong measure zero (i.e. X cannot be covered by $\{I_n\}_{n \geqslant n_0}$ of lengths ε_n).

Toward the proof of Lemma 8.2, we start with a property of Laver forcing:

8.3 Lemma

Let $p \Vdash \varphi_1 \vee \cdots \vee \varphi_k$. Then there is $q \leqslant_0 p$ such that

$$\exists i \leqslant k \quad q \Vdash \varphi_i \qquad (*)$$

Proof We use a fusion argument. Let us recall that for $t \in p$,

$$S(t) = \{a : t^\frown a \in p\}$$

Let $s = \text{stem}(p)$. Assume that the lemma fails. Then there are only finitely many $a \in S(s)$ such that $\exists q \leqslant_0 p \upharpoonright (\widehat{sa})$ with property $(*)$. By removing those finitely many nodes (and the nodes above them), we get $p_1 \leqslant_0 p$. For every $\widehat{sa} \in p_1$ there are only finitely many $b \in S(\widehat{sa})$ such that $\exists q \leqslant_0 p_1 \upharpoonright (\widehat{sab})$ with property $(*)$. By removing all such b we obtain $p_2 \leqslant_1 p$. Continuing in this way we get a fusion sequence $p \geqslant_0 p_1 \geqslant_1 p_2 \geqslant_2 \cdots$, and let $r = \bigcap_{n<\omega} p_n$. By the construction of the p_n's, if $t \in r$ then there is no $q \leqslant_0 r \upharpoonright t$ with property $(*)$. But then no $q \leqslant r$ forces $(\exists i \leqslant k)\, \varphi_i$, a contradiction. □

We need two more lemmas before we prove Lemma 8.2:

8.4 Lemma

Let p be a condition with stem s and let $\dot{x}$ be a name for a real (in $[0,1]$). Then there is a condition $q \leqslant_0 p$ and a real u such that for every $\varepsilon > 0$, for all but finitely many $a \in S(s)$,

$$q \upharpoonright (\widehat{sa}) \Vdash |\dot{x} - u| < \varepsilon$$

Proof Let $\{t_n\}_n$ be an enumeration of $\{\widehat{sa} : a \in S(s)\}$. For each n we pick, by Lemma 8.3, a condition $q_n \leqslant_0 p \upharpoonright t_n$ and an interval $J_n = [m/n, (m+1)/n]$ so that $q_n \Vdash \dot{x} \in J_n$. There is a sequence $\{k_n\}_{n<\omega}$ so that the J_{k_n} form a decreasing sequence converging to a unique real u. Let $q = \bigcup_{n=0}^{\infty} q_{k_n}$; it is easy to see that q works. □

8.5 Lemma

Let p be a condition with stem s and let $\{\dot{x}_n\}_n$ be a sequence of names for reals. Then there exists a condition $q \leqslant_0 p$, and a set of reals $\{u_t : t \in q, t \supseteq s\}$ such that for every $\varepsilon > 0$ and every $t \in q$, $t \supseteq s$, for all but finitely many $a \in S(t)$,

$$q \upharpoonright (\widehat{ta}) \Vdash |\dot{x}_k - u_t| < \varepsilon$$

where $k = \text{length}(t) - \text{length}(s)$.

Proof Using a fusion argument, by a repeated application of Lemma 8.4. First we get $p_1 \leqslant_0 p$ and u_s. Then for every immediate successor t of s in p_1 we get $q_t \leqslant_0 p_1 \upharpoonright t$, and u_t, and let $p_2 = \bigcup_t q_t$. And so on; we get a fusion sequence $p \geqslant_0 p_1 \geqslant_1 p_2 \geqslant_2, \ldots$, and let $q = \bigcap_{n<\omega} p_n$. □

Proof of Lemma 8.2

Let $X \in V$ be a subset of $[0,1]$, and let $p \in P$ be such that $p \Vdash X$ has strong measure zero. We show that X is countable. Let s be the stem of p, $|s| = n$. Let G be generic on P, and let f be the Laver real given by G.

Consider the sequence $\{\varepsilon_k\}_{k\geqslant n}$ defined by

$$\varepsilon_k = \frac{1}{f(k)} \qquad (k \geqslant n)$$

There exists a sequence of intervals $I_k, k \leqslant n$, of length ε_k, so that $X \subseteq \bigcup_{k=n}^{\infty} I_k$. For each $k \geqslant n$, let x_k be the center of I_k.

Let $q \leqslant_0 p$ be a condition obtained by Lemma 8.5 (applied to p and $\{\dot{x}_k\}_{k\geqslant n}$); let $\{u_t : t \in q\}$ be the reals from Lemma 8.5. We shall show that $X \subseteq \{u_t : t \in q\}$.

Let $v \notin \{u_t : t \in q\}$ and show that $v \notin X$. Since $p \Vdash X \subseteq \bigcup_{k\geqslant n} \dot{I}_k$, it suffices to find some $r \leqslant q$ such that $r \Vdash v \notin \dot{I}_k$, for all $k \geqslant n$. We construct $r \leqslant q$ by induction on the levels of q; at stage $k \geqslant n$ we guarantee that $r \Vdash v \notin \dot{I}_k$.

For example, the first step is as follows: let $\varepsilon = (1/2)\cdot|v - u_s|$. For all but finitely many $a \in S(s)$, $q \restriction (\widehat{sa}) \Vdash |\dot{x}_n - u_s| < \varepsilon$. Also, for each a, $q \restriction (sa) \Vdash \dot{f}(n) = a$, and so $q \restriction (sa) \Vdash \dot{\varepsilon}_n = 1/a$; thus, for all but finitely many a, $q \restriction (\widehat{sa}) \Vdash |\dot{x}_n - v| > \dot{\varepsilon}_n$, in other words $v \notin \dot{I}_n$. Thus, by removing finitely many immediate successors of s, we ensure that $r \Vdash v \notin \dot{I}_n$. We continue in this way to get $r \leqslant q$ such that $r \Vdash v \notin \bigcup_{k\geqslant n} \dot{I}_k$. □

8.6 Theorem (Laver)

Assume the GCH. Let P be the countable support iteration of $\{\dot{Q}_\beta\}$ such that for each $\beta < \omega_2$, $\Vdash_\beta \dot{Q}_\beta$ is the Laver forcing. In the generic extension $V[G]$, all strong measure zero sets are countable.

First we wish to show that the iteration preserves cardinals. Preservation of $\aleph_1$ is crucial: the forcing is neither ω-closed, nor does it have the c.c.c. We state (without proof) the key lemma:

8.7 Lemma

For every countable set $X \in V[G]$ of ordinals there is $Y \in V$, countable in V, such that $X \subseteq Y$.

We remark that Lemma 8.7 follows from the more general property of countable support iteration of proper forcing, which will be proved in Part III, Chapter 5.

Preservation of cardinals $\aleph_2$ and up follows from Corollary 7.10; but for what we need:

8.8 Lemma

Assume the GCH. For every $\alpha < \omega_2, p \restriction \alpha$ has a dense subset of size $\aleph_1$, and $V[G_\alpha] \Vdash 2^{\aleph_0} = \aleph_1$.

Again, we omit the proof and remark that the lemma holds for countable support iteration of proper forcings, provided that $\Vdash_\beta |\dot{Q}_\beta| = 2^{\aleph_0}$ for every $\beta < \omega_2$.

8.9 Corollary
P has the $\aleph_2$-chain condition.

Proof Corollary 7.10. □

8.10 Corollary
$V[G]$ satisfies $2^{\aleph_0} = \aleph_2$.

Proof Because of the chain condition, every real in $V[G]$ is in $V[G_\alpha]$ for some $\alpha < \omega_2$. Hence, $2^{\aleph_0} \leqslant \aleph_2$ by Lemma 8.8. On the other hand, each G_α produces a Laver real, hence $2^{\aleph_0} \geqslant \aleph_2$. □

8.11 Lemma
Every set of reals of size $\aleph_1$ in $V[G]$ is in $V[G_\alpha]$ for some $\alpha < \omega_2$.

Proof Follows from the chain condition. □

We shall now sketch a proof of Theorem 8.6. We have to show that no uncountable set of reals in $V[G]$ has strong measure zero. It suffices to show that if $|X| = \aleph_1$ then X does not have strong measure zero. Thus, let $X \in V[G]$ be a set of reals of size $\aleph_1$.

By Lemma 8.11, $X \in V[G_\alpha]$ for some $\alpha < \omega_2$. Now it follows from Lemma 8.2 that in $V[G_{\alpha+1}]$ there is a sequence $\{\varepsilon_n\}_n$ witnessing that X does not have strong measure zero. But that only means that X cannot be covered by intervals of length ε_n in $V[G_{\alpha+1}]$; we need, however, that X does not have strong measure zero in $V[G]$.

The Factor Lemma 7.13 reduces the problem. By the lemma, the iteration P is equivalent to $P_\alpha * \dot{Q}$, where $\dot{Q}$ is, in V^{P_α}, the countable support iteration of Laver forcings of length ω_2. Thus, the following lemma completes the proof:

8.12 Lemma
If X is an uncountable set of reals in V, then X does not have strong measure zero in $V[G]$.

In fact, if f is a Laver real over V, and $\varepsilon_n = 1/f(n)$ ($n < \omega$), then for some n_0, X cannot be covered in $V[G]$ by intervals $\{I_n\}_{n \geqslant n_0}$ of length ε_n. We omit the proof of Lemma 8.12; it is not unlike the proof of Lemma 8.2, but its fusion arguments are more delicate, as it involves conditions $p \in P$ rather than just Laver trees.

References

Laver, R. (1976). *Acta. Math.* **137**, 151–69.
Borel, E. (1919). *Bull. Soc. Math. France* **47**, 97–125.

PART III

Proper forcing

1 Stationary sets

We consider closed unbounded and stationary sets of countable ordinals. We assume that the reader is familiar with closed unbounded and stationary subsets of a regular uncountable cardinal κ. With the exception of Chapter 7, we only consider stationary subsets of κ for $\kappa = \aleph_1$; later in this chapter we generalize these notions. A *closed unbounded* subset of ω_1 (a *club*) is an unbounded set $C \subseteq \omega_1$ such that $\sup(C \cap \alpha) \in C$ for every $\alpha < \omega_1$. A set $S \subseteq \omega_1$ is *stationary* if $S \cap C \neq \emptyset$ for every club C.

For any set A, $A^{<\omega}$ denotes the set of all finite subsets of A. If F is a function on $A^{<\omega}$, we say that a nonempty $X \subseteq A$ is *closed under* F if $F(X^{<\omega}) \subseteq X$.

1.1 Proposition

(*a*) if $F: \omega_1{}^{<\omega} \to \omega_1$, then

$$C_F = \{\lambda < \omega_1 : \lambda \text{ is closed under } F\}$$

is a club.

(*b*) For every club C there is a function $F: \omega_1^{<\omega} \to \omega_1$ such that $C_F \subseteq C$. In fact, $C_F = C'$, the set of all limit points of C.

Proof

(*a*) Easy.

(*b*) Let $F(e) =$ least $\alpha \in C$ greater than $\max(e)$. Note that every $\lambda \in C_F$ is a limit ordinal. □

Preservation of stationary sets

If $V[G]$ is a generic extension, then every club $C \in V$ remains a club in $V[G]$, provided that $\aleph_1$ is preserved. But a stationary set S in V may no longer be stationary in $V[G]$, as there may be a club $C \in V[G]$ disjoint from S.

1.2 Example

Shooting a club through a given stationary set.

For any stationary set S there is a forcing notion P_S that generically adds a club C so that $C \subset S$, and preserves $\aleph_1$. If S is costationary, i.e. if $-S$ is also stationary, then $-S$ is destroyed by P_S as it is disjoint from C in $V[G]$.

Let S be a stationary set.

$$P_S = \text{the set of all (bounded) closed sets of ordinals } p \text{ so that } p \subset S$$
$$p \leqslant q \text{ if } p \text{ is an end-extension of } q \text{ (if } q = p \cap \alpha \text{ for some } \alpha) \quad (1.3)$$

Note that since p is closed, $\max(p)$ exists (a countable ordinal).

If G is a generic filter, let

$$C = \bigcup G$$

Clearly, C is a subset of S, and because, for every $\alpha < \omega_1$, the set $D = \{p : \max(p) \geqslant \alpha\}$ is dense in P_S, C is an unbounded set of ω_1. Also, $\operatorname{supp}(C \cap \alpha) \in C$ clearly holds for every $\alpha < \omega_1$. Thus, C is a club, provided that $\aleph_1$ is preserved. That follows from

1.4 Lemma

P_S is ω-distributive.

Proof We prove that every countable set of ordinals in $V[G]$ is in the ground model. Thus, let

$$p \Vdash \dot{f} : \omega \to \text{Ord}$$

we shall find a $q \leqslant p$ and some f so that $q \Vdash \dot{f} = f$.

By induction on α we construct a chain $\{A_\alpha\}_\alpha$ of countable subsets of P_S:

Let $A_0 = \{p\}$, and $A_\alpha = \bigcup_{\beta < \alpha} A_\beta$ when α is a limit. Given A_α, let $\gamma_\alpha = \sup\{\max(q) : q \in A_\alpha\}$. For each $q \in A_\alpha$ and each n, we choose some $r = r(q,n) \in P_S$ so that $r \leqslant q$, r decides $\dot{f}(n)$, and $\max(r) > \gamma_\alpha$. Then we let $A_{\alpha+1} = A_\alpha \cup \{r(q,n) : q \in A_\alpha, n < \omega\}$.

The sequence $\{\gamma_\alpha\}_\alpha$ is increasing and continuous. Let

$$C = \{\lambda : \text{if } \alpha < \lambda \text{ then } \gamma_\alpha < \lambda\}$$

As C is a club and S is stationary, there is a limit ordinal λ so that $\lambda \in C \cap S$. Let $\{\alpha_n\}_n$ be an increasing sequence with limit λ; then $\lim_n \gamma_{\alpha_n} = \lambda$ as well.

There is a sequence $\{p_n\}_n$ of conditions so that $p_0 = p$, and for every n, $p_{n+1} \in A_{\alpha_{n+1}}$, $p_{n+1} \leqslant p_n$, and p_{n+1} decides $\dot{f}(n)$. Since $\gamma_{\alpha_n} < \max(p_{n+1}) \leqslant \gamma_{\alpha_{n+1}}$, we have $\lim_n \max(p_n) = \lambda$, and because $\lambda \in S$, the closed set

$$q = \bigcup_n p_n \cup \{\lambda\}$$

is a condition. Since $q \leqslant p_n$ for each n, q decides each $f(n)$, and so there is some f such that $q \Vdash \dot{f} = f$. □

The following theorem gives sufficient conditions for a notion of forcing to preserve stationary sets:

1.5 Theorem

(*a*) If a notion of forcing P satisfies the countable chain condition, then every club $C \in V[G]$ has a club subset D such that $D \in V$. Consequently, every stationary set $S \in V$ remains stationary in $V[G]$.

(*b*) If P is ω-closed then every stationary set $S \in V$ remains stationary in $V[G]$.

Proof (*a*) Let

$$p \Vdash \dot{C} \text{ is a club}$$

and let $D = \{\alpha : p \Vdash \alpha \in \dot{C}\}$. Clearly, $p \Vdash D \subseteq \dot{C}$; we prove that D is a club. It is not difficult to see that D is closed; we shall use the c.c.c. to show that D is unbounded.

Let $\alpha_0 < \omega_1$ and let us find $\alpha > \alpha_0$ so that $p \Vdash \alpha \in \dot{C}$. There is a name $\dot{\gamma}$ for an ordinal so that

$$p \Vdash \alpha_0 < \dot{\gamma} \in \dot{C}$$

There is a partition W of p, and ordinals $\{\gamma_q : q \in W\}$ so that $q \Vdash \dot{\gamma} = \gamma_q$ for each $q \in W$. By the c.c.c., W is countable and so $\alpha_1 = \sup\{\gamma_q : q \in W\}$ is countable. Thus,

$$p \Vdash (\exists \gamma \in \dot{C}) \alpha_0 < \gamma \leqslant \alpha_1$$

Similarly, we find a sequence $\alpha_1 < \alpha_2 < \cdots$ so that, for every n,

$$p \Vdash (\exists \gamma \in \dot{C}) \alpha_n < \gamma \leqslant \alpha_{n+1}$$

Then we let $\alpha = \lim_n \alpha_n$, and it follows that

$$p \Vdash \alpha_0 < \alpha \in \dot{C}$$

(*b*) Let S be stationary, and let

$$p \Vdash \dot{C} \text{ is a club}$$

We shall find a $q \leqslant p$ so that $q \Vdash S \cap \dot{C} \neq \emptyset$.

We construct an increasing continuous ordinal sequence $\{\gamma_\alpha\}_\alpha$ and a decreasing sequence $\{p_\alpha\}_\alpha$ of conditions as follows.

Let $p_0 \leqslant p$ and γ_0 be so that $p_0 \Vdash \gamma_0 \in \dot{C}$. Given p_α and γ_α, we let

$p_{\alpha+1} \leqslant p_\alpha$ and $\gamma_{\alpha+1} > \gamma_\alpha$ be so that $p_{\alpha+1} \Vdash \gamma_{\alpha+1} \in \dot{C}$. When α is a limit, let p_α be a lower bound of $\{p_\beta\}_{\beta<\alpha}$ and $\gamma_\alpha = \sup\{\gamma_\beta\}_{\beta<\alpha}$; clearly $p_\alpha \Vdash \gamma_\alpha \in \dot{C}$.

Since $\{\gamma_\alpha\}_\alpha$ is increasing and continuous, and S is stationary, there is some α so that $\gamma_\alpha \in S$. Then

$$p_\alpha \Vdash \gamma_\alpha \in \dot{C} \cap S \qquad \square$$

Closed unbounded and stationary subsets of $[A]^\omega$

Let A be an uncountable set. By

$$[A]^\omega$$

we denote the set all (at most) countable subsets of A.

1.6 Definition

A set $C \subseteq [A]^\omega$ is *unbounded* if for every $x \in [A]^\omega$ there is $y \in C$ such that $x \subseteq y$. C is *closed* if for every chain

$$x_0 \subseteq x_1 \subseteq \cdots \subseteq x_n \subseteq \cdots \quad (n < \omega)$$

in C, the union $\bigcup_{n=0}^\infty x_n$ is in C. C is *closed unbounded* (*club*) in $[A]^\omega$ if it is closed and unbounded. A set $S \subseteq [A]^\omega$ is *stationary* (in $[A]^\omega$) if $S \cap C \neq \varnothing$ for every club C in $[A]^\omega$.

If A is a set of cardinality $\aleph_1$ then as the following easy fact shows, the concept of club and stationary coincides essentially with the usual concept of club and stationary (note that $\omega_1 \subset [\omega_1]^\omega$):

1.7 Proposition

The set ω_1 is a club in $[\omega_1]^\omega$. Moreover, a set $C \subseteq \omega_1$ is a club in $[\omega_1]^\omega$ iff it is a club subset of ω_1. A set $S \subseteq [\omega_1]^\omega$ is stationary iff $S \cap \omega_1$ is a stationary subset of ω_1; in particular, every stationary subset of ω_1 is stationary in $[\omega_1]^\omega$. $\square$

The closed unbounded subsets of $[A]^\omega$ generate a normal countably closed filter:

1.8 Theorem

(*a*) If $\{C_n\}_n$ are clubs then $\bigcap_{n=0}^\infty C_n$ is a club.

(*b*) The diagonal intersection

$$\Delta\{C_a : a \in A\} = \{x \in [A]^\omega : x \in \bigcap_{a \in x} C_a\}$$

of clubs $\{C_a\}_{a \in A}$ is a club.

(*c*) Every choice function

$$f(x) \in x$$

on a stationary set S is constant on a stationary subset $T \subseteq S$, i.e. there is $a \in A$ such that

$$f(x) = a$$

for all $x \in T$.

Proof (*a*) First we show that the intersection of two clubs is a club. Let $C = C_1 \cap C_2$, where C_1 and C_2 are clubs. C is clearly closed. In order to show that C is unbounded, let $x \in [A]^\omega$. We consider a chain $x \subseteq x_0 \subseteq y_0 \subseteq x_1 \subseteq y_1 \subseteq \cdots$, where each x_n is in C_1 and each y_n is in C_2. Then $y = \bigcap_{n=0}^\infty x_n = \bigcap_{n=0}^\infty y_n$ is in $C_1 \cap C_2$.

In order to prove (*a*) it suffices now to consider a descending sequence $C_0 \supseteq C_1 \supseteq \cdots \supseteq C_n \supseteq \cdots$ of clubs. Their intersection C is certainly closed. To show that C is unbounded, let $x \in [A]^\omega$. We consider a chain $x_0 \subseteq x_1 \subseteq x_2 \subseteq \cdots$, where each x_n is a member of C_n. Then, $y = \bigcup_{n=0}^\infty x_n$ belongs to each $C_k, k < \omega$ (because $y = \bigcup_{n=k}^\infty x_n$ is the union of a chain in C_k).

(*b*) Let $C_a, a \in A$, be clubs, and let

$$x \in C \quad \text{iff} \quad (\forall a \in x) x \in C_a$$

First we show that C is closed. Let $x_0 \subseteq x_1 \subseteq \cdots \subseteq x_n \subseteq \cdots$ be a chain in C, and let $x = \bigcup_{n=0}^\infty x_n$. To show that $x \in C$, let $a \in x$ and let us show that $x \in C_a$. There is k such that $a \in x_n$ for all $n \geqslant k$; hence, $x_n \in C_a$ for all $n \geqslant k$, and so $x \in C_a$.

Now we show that C is unbounded. Let $x_0 \in [A]^\omega$. By induction, let $x_0 \subseteq x_1 \subseteq \cdots \subseteq x_n \subseteq \cdots$ be as follows: given x_n, for every $a \in x_n$ choose $z(n,a) \in C_a$ such that $x_n \subseteq z(n,a)$, and let $x_{n+1} = \bigcup \{z(n,a) : a \in x_n\}$. Let $y = \bigcup_{n=0}^\infty x_n$, and let us show that $y \in C$.

Thus, let $a \in y$ and we show that $y \in C_a$. The sequence $\{z(n,a) : n < \omega\}$ is a chain in C_a, because

$$x_0 \subseteq z(0,a) \subseteq x_1 \subseteq \cdots \subseteq z(n,a) \subseteq \cdots$$

and so $y = \bigcup_{n=0}^\infty z(n,a)$ belongs to C_a.

(*c*) Let S be stationary, and let f be a choice function on S.

In order to show that f is constant on a stationary subset of S, let us assume that, on the contrary, for every $a \in A$ there is a club C_a disjoint from S such that $f(x) \neq a$ on C_a. Let C be the diagonal intersection of $\{C_a : a \in A\}$. Since S is stationary, and C is a club, there is some $x \in S$ such that $x\in C_a$ for all $a\in x$. Hence, $f(x)\neq a$ for all $a\in x$, contrary to the assumption that $f(x)\in x$. □

By the following lemma, the condition of closure under union of chains in Definition 1.6 can be replaced by unions of directed sets. A set $D\subseteq[A]^\omega$ is *directed* if for every $x, y\in D$ there is a $z\in D$ with $x\cup y\subseteq z$.

1.9 Lemma

Let $C\subseteq[A]^\omega$. If C is closed then for every countable directed $D\subseteq C$, $\bigcup D\in C$.

Proof Let $D=\{x_n\}_n$; we construct a chain $\{y_n\}_n$ in C with the property that $\bigcup_{n=0}^{\infty}y_n=\bigcup D$. Let $y_0=x_0$, and for each n let $y_{n+1}=x_k$, where k is the least k such that $x_k\supseteq y_n\cup x_{n+1}$. □

We now characterize closed unbounded and stationary sets in terms of functions $F:A^{<\omega}\to A$. As in Proposition 1.1(*a*), if $F:A^{<\omega}\to A$ then the set

$$C_F=\{x\in[A]^\omega : x \text{ is closed under } F\}$$

is a club. More generally

1.10 Proposition

For every $F:A^{<\omega}\to[A]^\omega$, the set

$$C_F=\{x\in[A]^\omega:(\forall e\in x^{<\omega})F(e)\subseteq x\}$$

is a club.

Proof Easy. □

1.11 Theorem

For every club C in $[A]^\omega$ there is a function $F:A^{<\omega}\to A$ such that C contains the club C_F.

Corollary A set S is stationary in $[A]^\omega$ if and only if for every function $F:A^{<\omega}\to A$ there is $x\in S$ closed under F.

Proof Let C be a club in $[A]^\omega$, and let us assume, w.l.o.g., that A is a cardinal, $A=\lambda$. By induction on the size of $e\in A^{<\omega}$, we construct a function $G:A^{<\omega}\to C$ such that for every $e\in A^{<\omega}$, $e\subseteq G(e)$, and that $G(e')\subseteq G(e)$ for all $e'\subseteq e$.

As each $G(e)$ is countable, we enumerate it by ω, and obtain countably many functions $G_k:A^{<\omega}\to A$, $k<\omega$, so that for each e, $G(e)=\{G_k(e):k<\omega\}$. Let us fix a pairing function $n\to(k_n,m_n)$ and define $F:A^{<\omega}\to A$ as follows: for each $\alpha\in A$, let $F(\{\alpha\})=\alpha+1$. For $e=\{\alpha_1<\cdots<\alpha_n\}$, let $F(e)=G_{k_n}(\{\alpha_1,\ldots,\alpha_{m_n}\})$.

We claim that $C_F\subseteq C$. Thus, let $x\in[A]^\omega$ be such that $F(x^{<\omega})\subseteq x$, and let us prove that $x\in C$. Let $e\in x^{<\omega}$, $e=\{\alpha_1<\cdots<\alpha_m\}$. We claim that $G(e)\subseteq x$. To show that, we need $G_k(e)\in x$ for every k. Let $n\geqslant m$ be such that $k=k_n$, $m=m_n$. The order type of x is a limit ordinal (because $F(\{\alpha\})=\alpha+1$ for every $\alpha\in x$) and so there are $\alpha_{m+1},\ldots,\alpha_n$ in x such that $G_k(e)=F(\{\alpha_1,\ldots,\alpha_m,\ldots,\alpha_n\})\in x$.

Hence, $x=\bigcup\{G(e):e\in x^{<\omega}\}$; in other words the union of a countable directed subset of C. Therefore $x\in C$. □

Let $A\subset B$ be uncountable sets. For every $X\subseteq[A]^\omega$, let

$$X^*=\{x\in[B]^\omega:x\cap A\in X\} \tag{1.12}$$

For every $Y\subseteq[B]^\omega$, let

$$Y\restriction A=\{x\cap A:x\in Y\} \tag{1.13}$$

1.14 Theorem

Let $A\subseteq B$.

(*a*) If C is a club in $[A]^\omega$ then C^* is a club in $[B]^\omega$. If S is stationary in $[B]^\omega$ then $S\restriction A$ is stationary in $[A]^\omega$.

(*b*) If C is a club in $[B]^\omega$ then $C\restriction A$ contains a club in $[A]^\omega$. If S is stationary in $[A]^\omega$ then S^* is stationary in $[B]^\omega$.

Proof (*a*) is easy. For (*b*), it suffices to prove the statement on clubs. So let C be a club in $[B]^\omega$. By Theorem 1.11 there is $F:B^{<\omega}\to B$ such that $C\supseteq C_F$. For each $e\in B^{<\omega}$, let $G(e)$ be the closure of e under F (the smallest $x\supseteq e$ closed under F). The set $C_G=\{x\in[B]^\omega:G(e)\subseteq x \text{ for all } e\in x^{<\omega}\}$ is a club and $C_G\subseteq C_F$. Let

$$D=\{x\in[A]^\omega:(\forall e\in x^{<\omega})G(e)\cap A\subseteq x\}$$

The set D is clearly a club. For every $x\in D$, let y be the closure of x under F. We have $y\in C_G$ and $y\cap A=x$. Hence, $x\in C_G\restriction A$ and so $D\subseteq C\restriction A$. □

The following is a useful proposition. We recall that if $\mathcal{M}$ is a model (for a given first order language) then the set of all elementary submodels of $\mathcal{M}$ is closed under unions of chains. Thus:

1.15 Proposition

Let $\mathcal{M} = \langle M, \ldots \rangle$ be an uncountable model. The set of all countable elementary submodels of $\mathcal{M}$ is a club in $[M]^\omega$. □

Preservation of stationary sets

We now re-examine Theorem 1.5.

1.16 Theorem

(*a*) If a notion of forcing P satisfies the countable chain condition, then for every uncountable A, every club $C \in V[G]$ in $[A]^\omega$ has a club subset D such that $D \in V$. Consequently, every stationary set $S \in V$ in $[A]^\omega$ is a stationary set in $[A]^\omega$ in $V[G]$.

(*b*) If P is ω-closed then every stationary $S \in V$ in $[A]^\omega$ remains stationary in $V[G]$.

(Note that $[A]^\omega$ is possibly a larger set in $V[G]$ then $[A]^\omega$ in V.)

Proof (*a*) Let

$$p \Vdash \dot{C} \text{ is a club}$$

There is a name $\dot{F}$ for a function: $A^{<\omega} \to A$ such that

$$p \Vdash C_{\dot{F}} \subseteq \dot{C}$$

Let $G: A^{<\omega} \to [A]^\omega$ be the function defined by

$$G(e) = \{a \in A: p \cdot \| \dot{F}(e) = a \| \neq 0\}$$

($G(e)$ is countable because P satisfies the c.c.c.) We claim that

$$p \Vdash C_G \subseteq \dot{C} \tag{1.17}$$

We have $p \Vdash \dot{F}(e) \in G(e)$. Thus, if x is such that $\forall e \in x^{<\omega} G(e) \subseteq x$, then $p \Vdash \dot{F}(e) \in x$; therefore $p \Vdash C_G \subseteq C_{\dot{F}}$, and (1.17) follows.

(*b*) Let S be a stationary set in $[A]^\omega$. In order to show that S remains stationary in $V[G]$, let

$$p \Vdash \dot{F}: A^{<\omega} \to A$$

It suffices to find a condition $q \leqslant p$ and some $x \in S$ such that

$$q \Vdash \dot{F}(x^{<\omega}) \subseteq x \tag{1.18}$$

Consider the model

$$\mathcal{M} = \langle V_\kappa, \in, (P, <), p, \dot{F}, \Vdash \rangle$$

where κ is a sufficiently large cardinal and $A \subseteq V_\kappa$. Let C be the set of all countable elementary submodels in $\mathcal{M}$; C is a club in $[V_\kappa]^\omega$.

By Theorem 1.14, there is $N \prec \mathcal{M}$ such that $N \cap A \in S$. We claim that $x = N \cap A$ satisfies (1.18). Enumerate $x^{<\omega} = \{e_n : n < \omega\}$, and construct a descending sequence of conditions $p = p_0 \geqslant p_1 \geqslant \cdots \geqslant p_n \geqslant \cdots$ such that for each n there is some $a_n \in N \cap A$ with

$$p_n \Vdash \dot{F}(e_n) = a_n$$

Such a_n exists in N because N is an elementary submodel of $\mathcal{M}$. Then let q be a lower bound for $\{p_n\}_n$. Now (1.18) follows. □

References

Jech, T. (1973). *Ann. Math. Logic* **5**, 165–98.
Kueker, D. (1977). *Ann. Math. Logic* **11**, 57–103.
Shelah, S. (1982). *Proper Forcing*. Lecture Notes in Mathematics 940, Springer-Verlag, NY.

2 *Infinite games on complete Boolean algebras*

In this chapter we consider several properties of complete Boolean algebras, and of the corresponding forcing extensions, formulated in terms of infinite games. Throughout the chapter, let B be a complete Boolean algebra, let P be a (separative) partial ordering, dense in B, and $V^P = V^B$ (or $V[G]$) be the corresponding Boolean-valued (or generic) model.

First we consider the *descending chain game* $\mathscr{G}_0$. Two players, I and II, choose successively the members of a descending chain

$$a_0 \geqslant b_0 \geqslant a_1 \geqslant b_1 \geqslant \cdots \geqslant a_n \geqslant b_n \geqslant \cdots \tag{2.1}$$

of nonzero elements of B. I chooses the a_n's and II chooses the b_n's. I wins the game (2.1) if and only if the sequence converges to zero (and II wins iff the sequence (2.1) has a nonzero lower bound).

We shall concern ourselves with the existence of winning strategies in $\mathscr{G}_0$ and in the other games defined below. It is easy to see that if the game $\mathscr{G}_0$ is played on P instead of on B (i.e. the chain (2.1) is in P, and I wins if the chain has no lower bound) then I (or II) has a winning strategy in the game played on P if and only if he has a winning strategy in the game played on B. The same is true for the other games to be considered.

2.2 Proposition

I has a winning strategy in $\mathscr{G}_0$ if and only if B is not ω-distributive.

Proof Assume that I has a winning strategy σ. Let $a_0 = \sigma(\langle\ \rangle)$ be I's initial move. We shall find a sequence $\{W_n\}$ of partitions of a_0 that do not have a common refinement. Let W_0 be a maximal incompatible set of elements $a_1 \in B$ of the form $a_1 = \sigma(\langle a_0, b_0\rangle)$, where $a_0 \geqslant b_0$. In general, let W_n be a maximal incompatible set of $a_{n+1} \in B$ such that $a_{n+1} = \sigma(\langle a_0, b_0, \ldots, a_n, b_n\rangle)$, where $a_0 \geqslant b_0 \geqslant \cdots \geqslant a_n \geqslant b_n$, and that the a_k's are chosen according to σ. Each W_n is a partition of a_0, and since σ is a winning strategy for I, each chain through the W_n converges to zero.

Conversely, assume that B is not ω-distributive, and let $\{D_n\}_n$ be a sequence of sets open dense below some a_0 such that $\bigcap_{n=0}^{\infty} D_n$ is empty. We define σ as follows: let $\sigma(\langle\ \rangle) = a_0$, and let $\sigma(\langle a_0, b_0, \ldots, a_n, b_n\rangle) \in D_n$. Clearly σ is a winning strategy for I. □

It is clear that if P is ω-closed then II has a winning strategy in the game $\mathscr{G}_0$. By a theorem of Foreman, the converse is true if $|P| \leqslant \aleph_1$. (Precisely: if II has a winning strategy and $|P| \leqslant \aleph_1$ then P has a σ-closed dense subset.) It is an open problem whether the converse is true in general.

One consequence of Foreman's result is the following equivalent of the existence of a winning strategy for II ([Jech, 1984], Addenda):

2.3 Proposition

II has a winning strategy in $\mathscr{G}_0$ on P if and only if there is a Q such that $P \times Q$ has a dense ω-closed subset. □

There is a B such that neither player has a winning strategy in $\mathscr{G}_0$. An example is the forcing notion P_S from Example 1.2 (if S is costationary). Since P_S is ω-distributive, I does not win. We omit the proof that II does not win, since we show in Chapter 3 that II does not win in the game $\mathscr{G}$ on P_S which is easier for II than $\mathscr{G}_0$. (All the games considered here are undetermined in general.)

Next we consider the *cut and choose* game $\mathscr{G}_1$. Player I begins by selecting some $p \in B$ and a partition W_0 of p. II responds by choosing some $u_0 \in W_0$. At the nth move, I plays a partition W_n of p and II chooses $u_n \in W_n$:

$$W_0, u_0, W_1, u_1, \dots, W_n, u_n, \dots \tag{2.4}$$

II wins the game $\mathscr{G}_1$ if and only if the sequence $\{u_n\}_n$ has a lower bound, i.e. iff

$$\exists q \leqslant p \quad \forall n \quad q \leqslant u_n \tag{2.5}$$

2.6 Proposition (Jech and Veličković)

The game $\mathscr{G}_0$ and $\mathscr{G}_1$ are equivalent, i.e. I (II) has a winning strategy in one iff I (II) has a winning strategy in the other.

Proof It is not difficult to prove that I has a winning strategy in $\mathscr{G}_1$ if and only if B is not ω-distributive. Thus it suffices to show that II has a winning strategy in $\mathscr{G}_0$ iff II has a winning strategy in $\mathscr{G}_1$.

Assume that II has a winning strategy σ in $\mathscr{G}_0$. We shall describe a strategy τ for II in $\mathscr{G}_1$. Playing $\mathscr{G}_1$, player I chooses P and plays W_0. Let $a_0 = p$ and apply σ as if I plays a_0 in $\mathscr{G}_0$; i.e. let $b_0 = \sigma(a_0)$. There is $u_0 \in W_0$ such that $u_0 \cdot b_0 \neq 0$. Let $\tau(W_0) = u_0$. When I plays $\mathscr{G}_1$, let $a_1 = u_0 \cdot b_0$, let $b_1 = \sigma(a_0, b_0, a_1)$, and let $\tau(W_0, u_0, W_1)$ be some $u_1 \in W_1$

so that $u_1 \cdot b_1 \neq 0$. In general, let $\tau(W_0, u_0, \ldots, W_n) = u_n \in W_n$ be so that $u_n \cdot b_n \neq 0$, where $b_n = \sigma(a_0, b_0, \ldots, a_n)$, $a_n = u_{n-1} \cdot b_{n-1}$. The strategy τ is a winning strategy for II in $\mathscr{G}_1$.

Conversely, assume that II has a winning strategy τ in $\mathscr{G}_1$. We shall describe a strategy σ for II in $\mathscr{G}_0$. Playing $\mathscr{G}_0$, I chooses some a_0. We observe that there is $b_0 \leqslant a_0$ with the property that for all $u \leqslant b_0$

$$\exists W \text{ partition of } a_0 \text{ such that } u = \tau(a_0; W) \tag{2.7}$$

(Otherwise, the set of all $u \leqslant p$ for which (2.7) fails is dense below p, and so there is a partition Z of p such that (2.7) fails for each $u \in Z$; a contradiction.) Thus, we (player II) choose such b_0, and let $\sigma(a_0) = b_0$. When I plays $a_1 \leqslant b_0$, let W_0 be a partition of a_0 such that $a_1 = \tau(a_0; W_0)$. By a similar argument, there is $b_1 \leqslant a_1$ with the property that for all $u \leqslant b_1$,

$$\exists W \text{ partition of } a_1 \text{ such that } u = \tau(a_0; W_0, W_1)$$

We let $\sigma(a_0, a_1) = b_1$. In general, at move n, we let $\sigma(a_0, \ldots, a_n)$ be some b_n so that $\forall u \leqslant b_n \exists W$ partition of a_n with $u = \tau(W_0, \ldots, W_{n-1}, W)$, and when I plays a_{n+1}, we find W_n so that $a_{n+1} = \tau(W_0, \ldots, W_n)$. The strategy σ is a winning strategy for II in $\mathscr{G}_0$. □

There is another version of the game $\mathscr{G}_1$, involving the Boolean-valued model (the *forcing version* of $\mathscr{G}_1$) as follows.

Player I starts the game by selecting a condition p, and chooses a name $\dot{\alpha}_0$ for an ordinal. Player II responds by choosing an ordinal β_0. At the nth move, I plays a name for an ordinal, $\dot{\alpha}_n$, and then II plays an ordinal β_n:

$$p; \dot{\alpha}_0, \beta_0, \dot{\alpha}_1, \beta_1, \ldots, \dot{\alpha}_n, \beta_n, \ldots \tag{2.8}$$

II wins the game if and only if

$$\exists q \leqslant p \quad \forall n \quad q \Vdash \dot{\alpha}_n = \beta_n \tag{2.9}$$

Since ordinal names correspond to partitions, namely

$$W = \{p \cdot [\![\dot{\alpha} = \gamma]\!] : \gamma \text{ an ordinal}\}$$

this game is just a reformulation of $\mathscr{G}_1$.

Next we consider the *countable choice* game $\mathscr{G}_\omega$. Player I begins by selecting a condition P, and chooses an ordinal name $\dot{\alpha}_0$. Player II chooses a countable set of ordinals B_0. At the nth move, I plays an ordinal name $\dot{\alpha}_n$, and II plays a countable set B_n:

$$p; \dot{\alpha}_0, B_0, \dot{\alpha}_1, B_1, \ldots, \dot{\alpha}_n, B_n, \ldots \tag{2.10}$$

II wins the game $\mathscr{G}_\omega$ if and only if

$$\exists q \leqslant p \quad \forall n \quad q \Vdash \dot{\alpha}_n \in B_n \tag{2.11}$$

Clearly, $\mathscr{G}_\omega$ is an easier game to play for player II than $\mathscr{G}_1$. Thus, if II has a winning strategy in $\mathscr{G}_1$ then II has a winning strategy in $\mathscr{G}_\omega$, and, similarly, if I does not have a winning strategy in $\mathscr{G}_1$ (if B is ω-distributive) then I does not have a winning strategy in $\mathscr{G}_\omega$.

There are many examples of forcing notions, for which II wins the game $\mathscr{G}_\omega$; this will be discussed in Chapter 4. A sufficient condition for I to win $\mathscr{G}_\omega$ is when P collapses ω_1, or more generally when P adds a countable set A of ordinals that is not included in any countable set in the ground model. If that is the case, let p be a condition that forces it, and let $\{\dot{\alpha}_0, \dot{\alpha}_1, \ldots, \dot{\alpha}_n, \ldots\}$ be a sequence of names such that

$$p \Vdash \dot{A} = \{\dot{\alpha}_n\}_n \text{ cannot be covered by a set countable in } V \tag{2.12}$$

Then I has a winning strategy in $\mathscr{G}_\omega$, playing the $\dot{\alpha}_n$.

The game $\mathscr{G}_\omega$ has a version formulated in terms of partitions: I selects p and at move n plays a partition W_n of p; II plays a countable subset B_n of W_n:

$$p; W_0, B_0, \ldots, W_n, B_n, \ldots \tag{2.13}$$

II wins if and only if

$$\prod_{n=0}^{\infty} \sum \{u : u \in B_n\} \neq 0 \tag{2.14}$$

(or, equivalently,

$$\exists q \leqslant p \quad \forall n \quad B_n \text{ is predense below } q)$$

Finally, we consider the *proper game* $\mathscr{G}$ (due to Charles Gray). Player I begins by selecting a condition p, and chooses an ordinal name $\dot{\alpha}_0$. Player II chooses an ordinal β_0. At the nth move, I plays an ordinal name $\dot{\alpha}_n$, and II plays an ordinal β_n:

$$p; \dot{\alpha}_0, \beta_0, \dot{\alpha}_1, \beta_1, \ldots, \dot{\alpha}_n, \beta_n, \ldots \tag{2.15}$$

II wins the game $\mathscr{G}$ if and only if

$$\exists q \leqslant p \quad \forall n \quad q \Vdash \exists k \quad \dot{\alpha}_n = \beta_k \tag{2.16}$$

An equivalent version of the proper game is when I plays ordinal names $\dot{\alpha}_n$ and II plays countable sets B_n, and II wins just in case

$$\exists q \leqslant p \quad \forall n \quad q \Vdash \exists k \quad \dot{\alpha}_n \in B_k \tag{2.17}$$

To see that the two versions are equivalent, note this: if II has a winning strategy in (2.17), then because $B = \bigcup_{k=0}^{\infty} B_k$ is countable, it is not hard to construct the set B as $\{\beta_n : n < \omega\}$ in ω moves. Thus, II has a winning strategy in (2.16). Similarly, if II can win against any strategy for I in (2.17), II can accomplish the same in (2.16).

Considering the version (2.17) it is easy to see that the game $\mathscr{G}$ is easier for II than the game $\mathscr{G}_\omega$. Thus we have:

$$\text{II has a winning strategy in } \mathscr{G}_1 \Rightarrow \text{II has a winning strategy in } \mathscr{G}_\omega \Rightarrow \text{II has a winning strategy in } \mathscr{G} \tag{2.18}$$

and similar implications for 'I does not have a winning strategy'.

Note that if a forcing notion P has property (2.12) then I has a winning strategy in $\mathscr{G}$ as well. Consequently, if II has a winning strategy in the proper game then every countable set of ordinals in $V[G]$ is included in a set in V that is countable in V, and hence $\aleph_1$ is preserved.

In Chapter 3 we show that if P destroys a stationary set then II does not have a winning strategy in the proper game, and so Example 1.2 gives an example of an undetermined proper game.

The proper game $\mathscr{G}_\omega$ also has a version formulated in terms of partitions: I selects p and at move n plays a partition W_n of p; II responds by choosing countable sets $B_0^n \subseteq W_0, B_1^n \subseteq W_1, \ldots, B_n^n \subseteq W_n$. II wins if and only if

$$\exists q \leqslant p \quad \forall n \quad \bigcup_{k=n}^{\infty} B_n^k \text{ is predense below } q \tag{2.19}$$

We leave the proof of equivalence to the reader.

We conclude this chapter by giving a (typical) example of a forcing notion for which II has a winning strategy in $\mathscr{G}_\omega$.

2.20 Example

II wins $\mathscr{G}_\omega$ for the Sacks forcing.

Let $(P, <)$ be the Sacks forcing, (Part I, Section 3.4). The proof that II has a winning strategy in $\mathscr{G}_\omega$ is exactly like the proof of Lemma 3.5, Part I: let I play p and then ordinal names $\dot{\alpha}_n$. As in Lemma 3.5, Part I, we find successively conditions p_n and countable (in fact finite) sets A_n so that

$$p \geqslant_0 p_0 \geqslant_1 p_1 \geqslant_2 \cdots \geqslant_n p_n \geqslant_{n+1} \cdots \tag{2.21}$$

and that for all n,

$$p_n \Vdash \dot{\alpha}_n \in A_n \tag{2.22}$$

Since (2.21) is a fusion sequence, it has a lower bound q, and

$$\forall n \quad q \Vdash \dot{\alpha}_n \in A_n$$

This gives a winning strategy for player II.

References

Jech, T. (1984). *Ann. Pure and Appl. Logic* **26**, 11–29.
Shelah, S. (1982). *Proper Forcing.* Lecture Notes in Mathematics 940, Springer-Verlag, NY.

3 Proper forcing

Proper forcing was introduced by S. Shelah, to describe a wide class of forcing notions that can be iterated (with countable support) without collapsing $\aleph_1$. The class of proper forcing includes all ω-closed forcings, all c.c.c. forcings, as well as many standard notions of forcing that adjoin generic reals.

We give four equivalent definitions of proper forcing.

3.1 A forcing theoretic definition

A notion of forcing $(P, <)$ is *proper* if, for every uncountable set A, every stationary subset of $[A]^\omega$ remains stationary in the generic extension $V[G]$.

3.2 A game theoretic definition

A notion of forcing $(P, <)$ is *proper* if player II has a winning strategy in the proper game (2.15) for P.

3.3 A model theoretic definition

In order to state the definition, let $(P, <)$ be a notion of forcing. We say that λ is *sufficiently large* if λ is a cardinal, and the power set of P is in V_λ. For any sufficiently large λ, we consider the model

$$V_\lambda = (V_\lambda, \in, P, <) \tag{3.4}$$

We recall that the set of all countable elementary submodels $M \prec V_\lambda$ is a club subset of $[V_\lambda]^\omega$. (As P and $<$ are assumed to be in the language of M the restriction of $\Vdash$ to any such M is definable in M.)

3.5 Lemma

Let λ be sufficiently large, let $M \prec (V_\lambda, \in, P, <)$ and let q be a condition in P. The following are equivalent:

(i) For every predense $W \subset P$ if $W \in M$ then $W \cap M$ is predense below q.
(ii) For every ordinal name $\dot{\alpha}$, if $\dot{\alpha} \in M$, then

$$q \Vdash \dot{\alpha} \in M$$

i.e.

$$q \Vdash \exists \beta \in M \quad \dot{\alpha} = \beta$$

(or, in detail, $\forall r \leqslant q \; \exists s \leqslant r \; \exists$ ordinal $\beta \in M \; s \Vdash \dot{\alpha} = \beta$).
(iii) $q \Vdash \dot{G} \cap M$ is an M-generic filter on $P \cap M$

We say that

$$q \text{ is } (P, M)\text{-generic} \tag{3.6}$$

if it satisfies any of the equivalent conditions in Lemma 3.5.

Proof We sketch proofs of some of the implications and leave the rest as an exercise.

(i)→(ii): Let $\dot{\alpha}$ be an ordinal name, $\dot{\alpha} \in M$. The set

$$W = \{p : \exists \beta \quad p \Vdash \dot{\alpha} = \beta\}$$

is dense, and $W \in M$ (because $M \prec V_\lambda$). By (i), $W \cap M$ is predense below q. If $p \in W \cap M$, then $\exists \beta \; p \Vdash \dot{\alpha} = \beta$, and because M is an elementary submodel, we have $\exists \beta \in M \; p \Vdash \dot{\alpha} = \beta$. Thus

$$W \cap M = \{p \in M : \exists \beta \in M \quad p \Vdash \dot{\alpha} = \beta\}$$

is predense below q, and (ii) follows.

(i)→(iii): Let G be a V-generic filter on P such that $q \in G$; we show that $G \cap M$ meets every $W \in M$ that is predense in $P \cap M$. The statement '$W \in M$ is predense in $P \cap M$' is easily seen to be equivalent to $M \Vdash$ (W is predense in P). Since M is an elementary submodel, W is predense in P, and by (i) $W \cap M$ is predense below q. Thus, $W \cap M \cap G$ is nonempty. □

3.7 Definition

A notion of forcing $(P, <)$ is *proper* if for some sufficiently large λ there is a club set $C \subseteq [V_\lambda]^\omega$ of countable elementary submodels $M \prec (V_\lambda, \in, P, <)$ with the following property:

$$\forall p \in M \quad \exists q \leqslant p \quad q \text{ is } (P, M)\text{-generic} \tag{3.8}$$

(Below we prove that 'for some sufficiently large λ' can be replaced by 'for all sufficiently large λ' in Definition 3.7.)

3.9 A Boolean-algebraic definition

This definition of properness is a variation on distributive laws. W is a *matrix of partitions* of a condition p, if

$$W = \{\alpha_{\alpha\beta} : \alpha, \beta < \lambda\} \tag{3.10}$$

such that for every α, $\{a_{\alpha\beta}\}_{\beta<\lambda}$ is a partition of p (with some $a_{\alpha\beta}$ possibly equal to 0).

3.11 Definition

A complete Boolean algebra B is *proper* if for every $p>0$, every uncountable λ and every matrix of partition (3.10) there is a club $C\subseteq[\lambda]^\omega$ such that for every $x\in C$,

$$\prod_{\alpha\in x}\sum_{\beta\in x}a_{\alpha\beta}\neq 0 \tag{3.12}$$

3.13 Theorem

The four definitions of properness given above are all equivalent.

Proof We start by showing that the model theoretic Definition 3.7 implies the game theoretic property 3.2. Let us consider the version (2.19) of the proper game, i.e. I selects some p and plays partitions W_n while II responds by playing countable sets $B_0^n, B_1^n,\dots,B_n^n$. To find a winning strategy for II, let λ be sufficiently large and let $C\subseteq[V_\lambda]^\omega$ of countable models $M\prec V_\lambda$ that all satisfy (3.8). When I selects p and plays W_0, let us choose $M_0\in C$ so that $p\in M_0$ and $W_0\in M_0$. Let $B_0=W_0\cap M_0$ be the first move of player II. In general, when I plays W_n, we choose $M_n\in C$ so that $M_n\supseteq M_{n-1}$ and $W_n\in M_n$. We let $B_0^n=W_0\cap M_n$, $B_1^n=W_1\cap M_n,\dots,B_n^n=W_n\cap M_n$. Since $M_0\subseteq M_1\subseteq\cdots$ are all in C, their union M is also in C. There is $q\leqslant p$ that is (P,M)-generic.

For every n, $W_n\in M$, and so $W_n\cap M$ is predense below q. But $W_n\cap M=\bigcup_{k=n}^\infty(W_n\cap M_k)=\bigcup_{k=n}^\infty B_n^k$. Hence, (2.19) holds, and the strategy described above is a winning strategy for player II in the proper game.

Next we show that the game theoretic property of a complete Boolean algebra B implies the distributive law (Definition 3.11). Let $p>0$, and let $W=\{a_{\alpha\beta}\}$ be a partition matrix of p. For each α, let W_α be the partition $\{a_{\alpha\beta}:\beta<\lambda\}$. Let σ be a winning strategy for II in the proper game (version (2.19)).

We define a function $F:\lambda^{<\omega}\to[\lambda]^\omega$. If e is a finite subset of λ, let $F(e)$ be a countable subset of λ such that if $\alpha_0,\dots,\alpha_k$ is any enumeration of e, and if $\{B_i^j\}_{i\leqslant j}^{j\leqslant k}$ are the moves of player II using the strategy σ against $W_{\alpha_0},\dots,W_{\alpha_k}$, then for every $a_{\alpha_i\beta}\in B_i^j$, β is in $F(e)$. By Proposition 1.10, the set C of all $x\in[\lambda]^\omega$ such that $F(e)\subseteq x$ for all $e\in x^{<\omega}$, is a club. We claim that every $x\in C$ satisfies (3.12).

Let $x\in C$, and let $x=\{\alpha_n\}_{n=0}^\infty$. Consider the proper game where I selects p and plays $\{W_{\alpha_n}\}$. Let $\{B_n^k\}$ be II's moves using the winning strategy σ. It follows from the way we defined F that, for every n,

$$\sum_{\beta\in x} a_{\alpha_n\beta} \geqslant \sum\left\{a : a\in \bigcup_{k=n}^{\infty} B_n^k\right\}$$

And because σ is a winning strategy, there is some $q \leqslant p$ such that (2.19) holds. Such a q witnesses (3.12).

Next we show that if B satisfies Definition 3.11 then every stationary set remains stationary in the generic extension by B. Let λ be uncountable, and let $S \subseteq [\lambda]^\omega$ be stationary. To show that S is stationary in V^B, let $\dot{F} \in V^B$ be such that

$$p \Vdash \dot{F} : \lambda^{<\omega} \to \lambda$$

for some condition p. It suffices to find $q \leqslant p$ and some $x \in S$ such that $q \Vdash \dot{F}(x^{<\omega}) \subseteq x$.

Let $\{e_\alpha\}_{\alpha<\lambda}$ enumerate $\lambda^{<\omega}$. The set $D = \{x \in [\lambda]^\omega : \forall e \in x^{<\omega} \exists \alpha \in x\ e = e_\alpha\}$ is a club. For each $\alpha, \beta < \lambda$, let

$$a_{\alpha\beta} = p \cdot \| \dot{F}(e_\alpha) = \beta \|$$

The collection $\{a_{\alpha\beta}\}$ is a partition matrix for p, and so there is a club $C \subseteq [\lambda]^\omega$ such that every $x \in C$ satisfies (3.12). Let $x \in S \cap C \cap D$, and let

$$q = \prod_{\alpha\in x} \sum_{\beta\in x} a_{\alpha\beta}$$

Now it follows that $q \Vdash \forall e \in x^{<\omega} \dot{F}(e) \in x$.

Finally we show that if P preserves stationary sets, then P satisfies the model theoretic definition of properness; that for all sufficiently large λ there is a club C with property (3.8). Toward a contradiction, assume that for some sufficiently large λ there is a stationary set $S \subset [V_\lambda]^\omega$ of countable models $M \prec V_\lambda$ for which (3.8) fails.

For each $M \in S$ there is a condition $p \in M$ for which (3.8) fails; by normality (1.8(c)), we may assume that the same condition p is a counterexample for each $M \in S$. Thus, for every $q \leqslant p$ and every $M \in S$, q is not (P, M)-generic.

Let $V[G]$ be a generic model such that $G \ni p$. Every partition W of p (in V) meets G in a unique condition q_W. Let C be the club set (in $V[G]$)

$$C = \{M \prec (V_\lambda)^V : \text{ if } \quad W \in M \quad \text{then} \quad q_W \in M\}$$

Since S remains stationary in $V[G]$, there is some $M \in S \cap C$. For every partition W of p, if $W \in M$ then $q_W \in W \cap M$ and so

$$\sum \{q : q \in W \cap M\} \in G$$

By the genericity of G, we have

$$\prod_{W \in M} \sum \{q : q \in W \cap M\} \in G$$

In other words, there is $q \in G$ such that each $W \cap M$ is predense below q, that is q is (P, M)-generic. This is a contradiction since $M \in S$. □

Reference

Shelah, S. (1982). *Proper Forcing*. Lecture Notes in Mathematics 940, Springer-Verlag, NY.

4 *Examples of proper forcing*

One of the definitions of properness is that the forcing preserves stationary sets. It follows therefore from Theorem 1.16 that every c.c.c. notion of forcing is proper, and so is every ω-closed forcing. In fact, this is easy to see when one uses the game definition: if P is ω-closed, then II has a winning strategy in the proper game (even in the game $\mathscr{G}_0$). If P is c.c.c. then every partition is at most countable, and again II has a winning strategy in the proper game (say in the version (2.19)).

4.1 Proposition

If P is proper then every countable set of ordinals X in $V[G]$ is included in a set $A \in V$ such that A is a countable set in V.

Proof We have shown this in Chapter 2. For another proof, let $X \subset \lambda$. The set $S = [\lambda]^\omega$ (defined in V) remains stationary in $V[G]$, and therefore meets the club set (in $V[G]$) $\{A \in [\lambda]^\omega : A \supseteq X\}$. The proposition follows. □

The class of proper forcings contains many other partial orderings in addition to the ω-closed and the c.c.c. ones. Of particular interest are the examples of generic reals considered in Chapter 3 of Part I.

4.2 Theorem

Sacks forcing, Prikry–Silver forcing, Mathias forcing, Laver forcing and Grigorieff forcing are all proper.

We omit the proof for Grigorieff reals. The other four examples can be all shown proper in a uniform way. The following property of forcing was introduced by Baumgartner:

4.3 Axiom A

A notion of forcing P satisfies *Axiom* A if there is a collection $\{\leqslant_n\}_n$ of partial orderings of P such that

(i) $p \leqslant_0 q$ implies $p \leqslant q$;
(ii) $p \leqslant_{n+1} q$ implies $p \leqslant_n q$;

(iii) if $\{p_n\}_n$ is a sequence such that

$$p_0 \geqslant_0 p_1 \geqslant_1 \cdots p_n \geqslant_n \cdots$$

then there is a q such that $q \leqslant_n p_n$ for all n;

(iv) for every $p \in P$ and for every $n < \omega$, if W is predense below p then there is a $q \leqslant_n p$ and a countable $B \subseteq W$ that is predense below q.

Note that (iv) is equivalent to:

(iv′) $\forall p \in P\ \forall n\ \forall \dot{\alpha}\ \exists q \leqslant_n p\ \exists B$ countable such that

$$q \Vdash \dot{\alpha} \in B$$

It is easy to see that every ω-closed P satisfies Axiom A (with $\leqslant_n$ being $\leqslant$) and so does every c.c.c. P (with $\leqslant_n$ being $=$). As for Sacks, Prikry–Silver, Mathias and Laver forcings, the respective partial orders $\leqslant_n$ defined in Chapter 3, Part I satisfy (i)–(iv) in the definition of Axiom A. (See Lemmas 3.5, 3.12 and 3.20 in Part I for the property (iv′). Even though we have not explicitly found a $q \leqslant_n p$ for each n, the proofs of these lemmas can be easily so modified.) Thus, it remains to prove the following:

4.4 Lemma

If P satisfies Axiom A then player II has a winning strategy in the game $\mathscr{G}_\omega$ on P and so P is proper.

Proof Let I select $p_0 \in P$ and an ordinal name $\dot{\alpha}_0$. By (iv′) there exists a condition $p_1 \leqslant_0 p_0$ and a countable set B_0 such that $p_0 \Vdash \dot{\alpha}_0 \in B_0$. At the nth move, when I plays $\dot{\alpha}_n$, there exists $p_{n+1} \leqslant_n p_n$ and a countable set B_n with $p_n \Vdash \dot{\alpha}_n \in B_n$. By (iii) there exists a condition q stronger than all the p_n. This condition q verifies that II wins. □

We briefly mention an example of a notion of forcing that is proper but does not satisfy Axiom A.

4.5 Example (Baumgartner) Adding a club with finite conditions

A condition $p \in P$ is a finite function with $\mathrm{dom}(p) \subset \omega_1$, $\mathrm{ran}(p) \subset \omega_1$, with the property that there exists a normal (i.e. increasing and continuous) function $f:\omega_1 \to \omega_1$ such that $p \subset f$. A condition q is stronger than p if $q \supseteq p$. A generic filter G yields a normal function $f_G:\omega_1 \to \omega_1$.

The forcing notion P has the property that player II wins in the proper game $\mathscr{G}$, but I wins in the game $\mathscr{G}_\omega$. We sketch the proof of this.

For the proper game, we first observe that the countable ordinals of the form $\alpha = \omega^\beta$ (*indecomposable* ordinals, which form a club set C_0) have the property that for every condition $p \subset \alpha \times \alpha$, the function $p \cup \{(\alpha, \alpha)\}$ is also a condition. If W is an antichain, let

$$C(W) = \{\lambda \in C_0 : \forall \alpha < \lambda \, \exists \beta < \lambda \, \forall p \subset \alpha \times \alpha \, \exists q \subset \beta \times \beta (q \in W \text{ and } q | p)\}$$

$C(W)$ is a club.

Player II wins the proper game as follows: when I selects p_0 and plays a partition W_0 of p_0, II chooses $\lambda_0 \in C(W_0)$ so that $p_0 \subset \lambda_0 \times \lambda_0$ and plays $B_0^0 = \{p \in W_0 : p \subset \lambda_0 \times \lambda_0\}$. At the nth move, when I plays W_n, II chooses $\lambda_n > \lambda_{n-1}$ in $C(W_0) \cap \cdots \cap C(W_n)$ and plays $B_k^n = \{p \in W_k : p \subset \lambda_n \times \lambda_n\}$ for $k = 0, \ldots, n$. We leave to the reader the proof that II wins.

In the game $\mathcal{G}_\omega$, player I has a winning strategy: let $\dot{f}$ be the canonical name for the generic normal function f_G. The first move of player I is the ordinal name $\dot{f}(0)$. When II responds by playing a countable set $B_0 \subset \omega_1$, I chooses an indecomposable ordinal α_1 greater than B_0 and plays $\dot{f}(\alpha_1)$. The play continues in the obvious way, and we leave it again to the reader to verify that I wins.

Reference

For more examples and counterexamples, see

Baumgartner, J. (1984). In *Handbook of Set-theoretical Topology* (Kunen, K. and Vaughan, J.E., eds.), pp. 913–59, North-Holland, Amsterdam.

5 Iteration of proper forcing

In this chapter we prove that properness is preserved by countable support iteration. It follows that when adding iteratively generic reals by proper forcing (such as Sacks reals, Laver reals, etc.), $\aleph_1$ is preserved.

5.1 Theorem

Let P_γ be a countable support iteration of length γ, of $\{\dot{Q}_\beta\}_\beta$, such that for every $\beta<\gamma$, $\Vdash_{\bar{\beta}}\dot{Q}_\beta$ is proper. Then P_γ is proper.

We know already that the theorem is true for $\gamma=2$, because when P preserves stationary sets and $\Vdash_{\bar{P}}\dot{Q}$ preserves stationary sets, then $P*\dot{Q}$ preserves stationary sets. We shall nevertheless give another proof of this fact, since the following proof will be generalized to iterations of arbitrary length. The proof uses the game theoretic definition of properness.

5.2 Two step iteration

We assume that P is proper, and that $\Vdash\dot{Q}$ is proper. We shall show that player II has a winning strategy in the proper game on $P*\dot{Q}$.

Let σ be a winning strategy for II on P, and let $\dot{\tau}\in V^P$ be such that $\Vdash_{\bar{P}}\dot{\tau}$ is a winning strategy for II on $\dot{Q}$. We shall describe a strategy for II on $P*\dot{Q}$: Player I starts the proper game by selecting a condition $(p,\dot{q})\in P*\dot{Q}$, and an ordinal name $\dot{\alpha}_0\in V^{P*Q}$. We describe II's response, an ordinal γ_0. We intend to step inside the model V^P and apply $\dot{\tau}$. The $P*\dot{Q}$-name $\dot{\alpha}$ can be thought of as (or identified with) a P-name for a $\dot{Q}$-name for an ordinal. Thus, let us argue in V^P and apply II's strategy $\dot{\tau}$ when I plays, in the proper game on $\dot{Q}$, the condition $\dot{q}$ and the ordinal $\dot{Q}$-name $\dot{\alpha}_0$. Let $\dot{\beta}_0$ be II's move given by $\dot{\tau}$. Back in the ground model, we consider the proper game on P, and use σ to respond to I's move p, $\dot{\beta}_0$. Let $\gamma_0=\sigma(p,\dot{\beta}_0)$.

At the nth move, I plays a $P*\dot{Q}$-name $\dot{\alpha}_n$. In V^P, $\dot{\alpha}_n$ is a $\dot{Q}$-name, and we apply the strategy $\dot{\tau}$ to find an ordinal $\dot{\beta}_n$. In V, we consider $\dot{\beta}_n$ (a P-name) as the nth move of I in the proper game on P, and use σ to return an ordinal γ_n.

Since $\dot{\tau}$ is a winning strategy for II (in V^P), we have

$$p\Vdash_P(\exists q'\leqslant\dot{q}\quad q'\Vdash_{\dot{Q}}\forall n\exists m\quad \dot{\alpha}_n=\dot{\beta}_n) \tag{5.3}$$

Hence, there is a $\dot{q}'$ such that $p \Vdash \dot{q}' \leqslant \dot{q}$ and

$$(p, \dot{q}') \Vdash \forall n \exists m \quad \dot{\alpha}_n = \dot{\beta}_m \tag{5.4}$$

Since σ is a winning strategy for II, there is $p' \leqslant p$ such that

$$p' \Vdash \forall m \exists k \quad \dot{\beta}_m = \gamma_k \tag{5.5}$$

Putting (5.4) and (5.5) together, we get

$$(p', \dot{q}') \Vdash \forall n \exists k \quad \dot{\alpha}_n = \gamma_k$$

thus showing that the strategy in the proper game on $P * \dot{Q}$ that we described is a winning strategy. □

Our next step in the proof is to show that iterations of length ω preserve properness. We shall repeatedly use the following lemma on two step iterations:

5.6 Lemma

Let $\dot{\alpha}$ be a $P * \dot{Q}$-name for an ordinal, and let $(p, \dot{r}) \in P * \dot{Q}$. There exists a P-name $\dot{\beta}$ and some $\dot{s}$ such that $p \Vdash \dot{s} \leqslant \dot{r}$, and

$$(p, \dot{s}) \Vdash \dot{\alpha} = \dot{\beta} \tag{5.7}$$

Proof There exists a partition W of p such that for every $w \in W$ there exists $\beta(w)$ and $\dot{s}(w)$ with the property that $w \Vdash \dot{s}(w) \leqslant \dot{r}$ and $(w, \dot{s}(w)) \Vdash \dot{\alpha} = \beta(w)$. Let $\dot{s}$ and $\dot{\beta}$ be the P-names such that, for every $w \in W$,

$$\| \dot{s} = \dot{s}(w) \|_B = w, \quad \| \dot{\beta} = \beta(w) \| = w$$

Now (5.7) follows. □

5.8 Iteration of length ω

We consider an iteration of length ω of proper notions of forcing $\{\dot{Q}_n\}_{n=1}^{\infty}$. We shall assume that the $\dot{Q}_n$'s are complete Boolean algebras; we are justified to assume that by Lemma 7.7 in Part II. Let B_0 be the trivial algebra $\{0, 1\}$, and for each n let $B_{n+1} = B_n * \dot{Q}_n$. Let P be the iteration of length ω, of the $\{\dot{Q}_n\}_n$. Thus, conditions in P are sequences $P = \langle p(0), p(1), \ldots, p(n), \ldots \rangle$ such that for each n, $p \restriction n \Vdash p(n) \in \dot{Q}_n$. For each $n \geqslant 1$, let $P_n = \{p \restriction n : p \in P\}$. P_n is dense in B_n.

The assumption is that for each n, $\Vdash_{B_n} \dot{Q}_n$ is a proper complete Boolean algebra. Let $\mathscr{G}(\dot{Q}_n)$ denote the proper game on $\dot{Q}_n$ (in V^{B_n}); by the assumption there are $\dot{\sigma}_n \in V^{B_n}$ such that for all n

$$\Vdash_{B_n} \dot{\sigma}_n \text{ is a winning strategy for II in } \mathscr{G}(\dot{Q}_n) \tag{5.9}$$

We wish to show that II wins in the game $\mathscr{G}(P)$.

Player I starts the game $\mathscr{G}(P)$ by selecting a condition p and chooses a P-name $\dot{\alpha}_0$. Let us apply Lemma 5.6 to $P = P_1 * P_{1,\omega}$ (where $P_{1,\omega} = P \upharpoonright (1, \omega)$; see (7.11), Part II). Let $p_0 = p(0) \in Q_0$. There exist P_1-names $\dot{\alpha}_0^0$ and $\dot{s}_0$ such that

$$\langle p_0 \rangle \Vdash_{P_1} \dot{s}_0 \in P_{1,\omega} \quad \text{and} \quad \dot{s}_0 \leqslant p \upharpoonright [1, \omega)$$

and

$$(p_0, \dot{s}_0) \Vdash \dot{\alpha}_0 = \dot{\alpha}_0^0$$

We apply the strategy σ_0 in the game $\mathscr{G}(Q_0)$ against I's move $p_0, \dot{\alpha}_0^0$, and obtain an ordinal β_0.

Player I then plays a P-name $\dot{\alpha}_1$. Let $\dot{p}_1 \in V^{P_1}$ be a name for a condition in $\dot{Q}_1$ such that $p_0 \Vdash \dot{p}_1 = \dot{s}_0(1)$. We apply Lemma 5.6 to $P_2 * P_{2,\omega}$. There exist P_2-names $\dot{\alpha}_1^1$ and $\dot{s}_1$ such that

$$\langle p_0 \dot{p}_1 \rangle \Vdash_{P_2} \dot{s}_1 \in P_{2,\omega} \quad \text{and} \quad \dot{s}_1 \leqslant \dot{s}_0 \upharpoonright [2, \omega)$$

and

$$(\langle p_0 \dot{p}_1 \rangle, \dot{s}_1) \Vdash \dot{\alpha}_1 = \dot{\alpha}_1^1$$

The P_2-name $\dot{\alpha}_1^1$ can be identified with a P_1-name for a $\dot{Q}_1$-name. Inside the model V^{P_1}, we start the game $\mathscr{G}(\dot{Q}_1)$. Player I begins by selecting $\dot{p}_1$ and by playing $\dot{\alpha}_1^1$. Applying the strategy $\dot{\sigma}_1$, II responds by playing an ordinal $\dot{\alpha}_1^0$ (so $\dot{\alpha}_1^0$ is a P_1-name for an ordinal). Then we continue the game $\mathscr{G}(Q_0)$, whose first moves were $p_0, \dot{\alpha}_0^0$ and β_0. The second move of player I is $\dot{\alpha}_1^0$, to which II responds by playing an ordinal β_1.

The game continues as in Table 5.10.

At the nth move, I chooses a P-name $\dot{\alpha}_n$. Let $\dot{p}_n \in V^{P_n}$ be a name for a condition in $\dot{Q}_n$ such that $\langle p_0 \dot{p}_1 \ldots \dot{p}_{n-1} \rangle \Vdash \dot{p}_n = \dot{s}_{n-1}(n)$. We apply Lemma 5.6 to $P_{n+1} * P_{n+1,\omega}$. There exist P_{n+1}-names $\dot{\alpha}_n^n$ and $\dot{s}_n$ such that

$$\langle p_0 \dot{p}_1 \ldots \dot{p}_n \rangle \Vdash_{P_{n+1}} \dot{s}_n \in P_{n+1,\omega} \quad \text{and} \quad \dot{s}_n \leqslant \dot{s}_{n-1} \upharpoonright [n+1, \omega) \tag{5.11}$$

and

$$(\langle p_0 \dot{p}_1 \ldots \dot{p}_n \rangle, \dot{s}_n) \Vdash \dot{\alpha}_n = \dot{\alpha}_n^n \tag{5.12}$$

Table 5.10

			I		II
$\mathscr{G}(P)$	$\cdots$	$\mathscr{G}(\dot{Q}_2)$	$\mathscr{G}(\dot{Q}_1)$	$\mathscr{G}(\dot{Q}_0)$	$\mathscr{G}(P)$
$\dot{\alpha}_0$				$\dot{\alpha}_0^0$	β_0
$\dot{\alpha}_1$			$\dot{\alpha}_1^1$	$\dot{\alpha}_1^0$	β_1
$\dot{\alpha}_2$		$\dot{\alpha}_2^2$	$\dot{\alpha}_2^1$	$\dot{\alpha}_2^0$	β_2
$\cdots$	$\cdots$	$\cdots$	$\cdots$	$\cdots$	$\cdots$

In V^{P_n}, we start the game $\mathscr{G}(\dot{Q}_n)$ by letting I play $\dot{p}_n$ and $\dot{\alpha}_n^n$. Applying $\dot{\sigma}_n$, II responds by playing a P_n-name $\dot{\alpha}_n^{n-1}$. Then we continue $\mathscr{G}(\dot{Q}_{n-1})$ in $V^{P_{n-1}}$ by letting I play $\dot{\alpha}_n^{n-1}$, to which II responds (by $\dot{\sigma}_{n-1}$) by playing $\dot{\alpha}_n^{n-2}$, and so on; eventually we obtain (by σ_0 in $\mathscr{G}(Q_0)$) an ordinal β_n.

It remains to show that the strategy described above is a winning strategy for II in $\mathscr{G}(P)$. Since $\dot{\sigma}_n$ is a winning strategy, we have

$$\exists q_0 \in Q_0, q_0 \leqslant p_0 \text{ such that } q_0 \Vdash \forall m \exists k\ \dot{\alpha}_m^0 = \beta_k \tag{5.13}$$

and also, for every n,

$$\langle p_0 \ldots \dot{p}_{n-1} \rangle \Vdash_{P_n} \exists q \in \dot{Q}_n, q \leqslant \dot{p}_n \text{ such that } q \Vdash_{\dot{Q}_n} \forall m \geqslant n \exists k\ \ \dot{\alpha}_m^n = \dot{\alpha}_k^{n-1} \tag{5.14}$$

Consequently, there is a sequence $\langle \dot{q}_n \rangle_n \leqslant_p \langle \dot{p}_n \rangle_n$ such that for every n, $\dot{q}_n$ satisfies $\langle q_0 \ldots \dot{q}_{n-1} \rangle \Vdash_{P_n} (\dot{q}_n \Vdash_{Q_n} \forall m \geqslant n \exists k\ \ \dot{\alpha}_m^n = \dot{\alpha}_k^{n-1})$. We claim that $q = \langle \dot{q}_n \rangle_n$ satisfies

$$q \Vdash \forall n \exists k\ \ \dot{\alpha}_n = \beta_k \tag{5.15}$$

To verify (5.15), let $n < \omega$. Because of (5.12) it suffices to verify that $q \leqslant (\langle p_0 \ldots \dot{p}_n \rangle, \dot{s}_n)$. However, this follows from the construction of the sequence $\{\dot{p}_n\}$.

Thus, II has a winning strategy in $\mathscr{G}(P)$. □

5.15 The iteration in general

The generalization of the above proof from ω to arbitrary limit ordinals is not trivial. To illustrate the difficulties, let γ be an ordinal of cofinality ω, say $\gamma = \lim_n \gamma_n$. By induction, we assume that each iteration $\dot{R}_n = P_{\gamma_{n-1}\gamma_n}$ is proper. Then the proof given in 5.8 shows that the ω-iteration R of $\{\dot{R}_n\}_n$ is proper. However, the iteration R need not be in general equivalent to the γ-iteration P. Given a sequence $\langle \dot{r}_n \rangle_n \in R$, each $\dot{r}_n$ is forced to be a function in $P_{\gamma_{n-1}\gamma_n}$ with countable support; but there is in general no reason why there should exist a countable set $S \subset \gamma$ such that $\forall n$ the support of $\dot{r}_n$ is included in S (so that $\langle \dot{r}_n \rangle_n$ can be represented as a function in P, with countable support).

We prove Theorem 5.1 by induction on the length of iteration. Simultaneously, we prove the following generalization of the Factor Lemma 7.13, Part II.

5.17 The ω-Factor Lemma

Let $\{\gamma_n\}_n$ be an increasing sequence of ordinals with $\lim_n \gamma_n = \gamma$. Let P be an iteration of length γ of proper $\{\dot{Q}_\alpha\}$. Let $R_0 = P_{\gamma_0}$, and for each n let $\dot{R}_{n+1} = P_{\gamma_n\gamma_{n+1}}$, and let R be the (full) iteration of $\{\dot{R}_n\}_{n<\omega}$. Then $B(R) = B(P)$.

Before we start the proof of Theorems 5.1 and 5.17, we introduce yet another version of the proper game. Let I select a condition p as usual, but instead of playing single ordinal names $\dot{\alpha}_n$, let him play names $\dot{A}_n$ for countable set of ordinals. II plays ordinals β_n as before, and wins if $\exists q \leqslant p$ such that

$$q \Vdash \forall n \forall \alpha \in \dot{A}_n \exists k \quad \alpha = \beta_k \tag{5.18}$$

It is not difficult to see that this version (seemingly more difficult for II) is equivalent to the proper game.

To formulate the right induction hypothesis, let P_γ be an iteration of proper forcings, and consider the proper game $\mathscr{G}$ on P_γ. Let σ be a strategy for II. We say that σ is a *good winning strategy* if for every sequence of moves $p, \dot{A}_0, \ldots, \dot{A}_n, \ldots$ of player I, σ produces a sequence $\{\beta_n\}_n$ such that there exists a $q \leqslant p$ that satisfies (5.18), and, moreover.

$$\text{support}(q) \subseteq \{\beta_n\}_{n<\omega} \tag{5.19}$$

We shall prove Theorem 5.1 by induction on γ. The inductive hypothesis (somewhat stronger than 'P is proper') reads as follows:

$$\forall \eta < \gamma(\Vdash_{P_\eta} \text{II has a good winning strategy in the proper game on } P_{\eta\gamma}) \tag{5.20}$$

We prove (5.20) by induction on γ. If γ is a successor ordinal then (5.20) follows easily from the earlier proofs. Thus, we assume that γ is a limit ordinal. If cf $\gamma = \omega$, then we first prove Lemma 5.17. The proof of (5.20) incorporates the procedure used in the proof of Lemma 5.17.

5.21 Proof of the ω-Factor Lemma

Let $\gamma = \lim_n \gamma_n$, and let P_γ be an iteration of length γ, of proper $\{\dot{Q}_\xi\}_{\xi<\gamma}$. For each n, let $\dot{R}_n = P_{\gamma_{n-1}\gamma_n}$ (and $R_0 = P_{\gamma_0}$); let R be the ω-iteration of $\{\dot{R}_n\}$. We wish to show that $B(R) = B(P_\gamma)$. For any $p \in P_\gamma$, let $r = \langle r_n \rangle_{n<\omega}$, where $r_n = p \restriction [\gamma_{n-1}, \gamma_n)$. Thus, P_γ embeds in R; it suffices to show that P_γ embeds in R densely.

Thus, let $r = \langle \dot{r}_n \rangle_n$ be a condition in R. We wish to find $p \in P_\gamma$ so that $p \leqslant r$. By the inductive hypothesis (5.20), each $\dot{R}_n$ has a good winning strategy $\dot{\sigma}_n$. We use these good strategies to produce p.

We play the proper games $\mathscr{G}(\dot{R}_n)$, simultaneously for all n. The game $\mathscr{G}(\dot{R}_n)$ is initiated with the condition $\dot{r}_n$. The moves of player I are names for countable sets of ordinals; the moves of player II are according to the strategy $\dot{\sigma}_n$.

At step 0, we start $\mathscr{G}(R_0)$. Player I plays r_0 and a name $\dot{A}_0^0$ for the support of $\dot{r}_1$. Player II responds β_0. At step 1, we start $\mathscr{G}(\dot{R}_1)$ in V^{R_0}. Player I plays $\dot{r}_1$ and a name $\dot{A}_1^1$ for the support of $\dot{r}_2$. Player II responds $\dot{\alpha}_1^0$. Then we continue the game $\mathscr{G}(R_0)$: player I plays $\dot{\alpha}_1^0$, and II responds β_1. Generally at step n, we start $\mathscr{G}(\dot{R}_n)$ in $V^{R_0 * \cdots * R_{n-1}}$. Player I plays $\dot{r}_n$ and a name $\dot{A}_n^n$ for the support of $\dot{r}_{n+1}$. Player II responds $\dot{\alpha}_n^{n-1}$. Then, playing $\mathscr{G}(\dot{R}_{n-1})$, player I plays $\dot{\alpha}_n^{n-1}$ and II responds $\dot{\alpha}_n^{n-2}$. And so on, until II plays β_n in the game $\mathscr{G}(R_0)$.

Since the $\dot{\sigma}_n$ are good winning strategies, there exists a condition $q = \langle \dot{q}_n \rangle_n \in R$, stronger than $\langle \dot{r}_n \rangle_n$, such that for every n, the following is forced by $q \restriction n$:

$$\dot{q}_n \Vdash_{\dot{R}_n} \forall \alpha \text{ played by I } \exists \beta \text{ played by II such that } \alpha = \beta \tag{5.22}$$

and

$$\text{the support of } \dot{q}_n \text{ is included in the set of all ordinals played by II} \tag{5.23}$$

Let $S = \{\beta_n\}_{n<\omega}$. It follows from (5.22) and (5.23) that, for every n,

$$q \restriction n \Vdash \text{support}(\dot{q}_n) \subseteq S$$

We shall conclude the proof by constructing a condition $p \in P_\gamma$ so that $p = q$ (under the embedding of P in R). This we do by induction on $\xi < \gamma$: if $\xi \notin S$ we let $p(\xi) = 1$, and if $\xi \in S$, then we let $p(\xi)$ be the condition $\dot{t} \in \dot{Q}_\xi$ so that $p \restriction \xi \Vdash \dot{t} = \dot{q}_n(\xi)$, where n is the unique n for which $\gamma_{n-1} \leqslant \xi < \gamma_n$. For each n we have $p \restriction \gamma_{n-1} \Vdash p \restriction [\gamma_{n-1}, \gamma_n) = \dot{q}_n$ and so $p = q$. □

5.24 Proof of the inductive condition (5.20)

Let γ be a limit ordinal, and let P_γ be an iteration of proper $\{\dot{Q}_\xi\}_{\xi<\gamma}$.

We want to show that (5.20) is true for γ. It suffices to prove it for $\eta = 0$: when $\eta < \gamma$ is arbitrary, we do the same proof inside V^{P_η}, because by the Factor Lemma, $P_{\eta\gamma}$ is in V^{P_η} a countable support iteration (of length $\gamma - \eta$) of proper forcings, and by the induction hypothesis,

$$\Vdash_{P_\eta} \forall \bar{\eta} \forall \bar{\gamma} \text{ such that } \eta \leqslant \bar{\eta} < \bar{\gamma} < \gamma (\Vdash_{P_{\bar{\eta}}} \text{II has a g.w.s. on } P_{\bar{\eta}\bar{\gamma}})$$

where g.w.s. = good winning strategy.

Thus, we want to show that II has a good winning strategy in the proper game $\mathscr{G}(P_\gamma)$. We shall follow closely the proof of 5.8 (iteration of length ω). If $\text{cf}\,\gamma = \omega$, we choose beforehand a cofinal sequence $\{\gamma_n\}_n$ with limit γ; if $\text{cf}\,\gamma > \omega$ then we define a certain increasing sequence $\{\gamma_n\}_n$ below γ in the course of the proof.

Player I starts the game $\mathscr{G}(P_\gamma)$ by selecting $p \in P_\gamma$ and by choosing a P_γ-name $\dot{\alpha}_0$. If cf $\gamma > \omega$, we choose some $\gamma_0 < \gamma$ (if cf $\gamma = \omega$, γ_0 is already given), and consider $R_0 = P_{\gamma_0}$. Let $p_0 = p \restriction \gamma_0$; by Lemma 5.6 there are R_0-names $\dot{\alpha}_0^0$ and $\dot{s}_0$ such that $p_0 \Vdash_{R_0} \dot{s}_0 \leqslant p \restriction [\gamma_0, \gamma)$ and $(p_0, \dot{s}_0) \Vdash \dot{\alpha}_0^0 = \dot{\alpha}_0$. We start the game $\mathscr{G}(R_0)$ by letting I play p_0 and an R_0-name for the countable set $\{\dot{\alpha}_0^0\} \cup \text{support}(\dot{s}_0)$. Player II uses a good winning strategy σ_0 to return β_0. We let this β_0 be player II's first move in $\mathscr{G}(P_\gamma)$.

At the nth move, I chooses a P_γ-name $\dot{\alpha}_n$. If cf $\gamma > \omega$, we choose some $\gamma_n < \gamma$, $\gamma_n > \gamma_{n-1}$, so that $\gamma_n > \beta_{n-1}$. Let $\dot{R}_n = P_{\gamma_{n-1}, \gamma_n}$. Let $\dot{p}_n$ be (a name for) a condition in $\dot{R}_n$ so that $\langle p_0 \ldots \dot{p}_{n-1} \rangle \Vdash \dot{p}_n = \dot{s}_{n-1} \restriction [\gamma_{n-1}, \gamma_n)$. By Lemma 5.6, there are $R_0 * \cdots * \dot{R}_n$-names $\dot{\alpha}_n^n$ and $\dot{s}_n$ such that

$$\langle p_0 \ldots \dot{p}_n \rangle \Vdash \dot{s}_n \in P_{\gamma_n \gamma} \quad \text{and} \quad \dot{s}_n \leqslant \dot{s}_{n-1} \restriction [\gamma_n, \gamma)$$

and

$$(\langle p_0 \ldots \dot{p}_n \rangle, \dot{s}_n) \Vdash \dot{\alpha}_n = \dot{\alpha}_n^n$$

We start the game $\mathscr{G}(\dot{R}_n)$ by letting I play $\dot{p}_n$ and $\dot{A}_n^n$, where $\dot{A}_n^n = \{\dot{\alpha}_n^n\} \cup \text{support}(\dot{s}_n)$.

II uses a good winning strategy $\dot{\sigma}_n$ to play $\dot{\alpha}_n^{n-1}$. Then we continue $\mathscr{G}(\dot{R}_{n-1})$ by letting I play $\dot{\alpha}_n^{n-1}$, to which II responds $\dot{\alpha}_n^{n-2}$. And so on, until II plays (by σ_0 in $\mathscr{G}(R_0)$) an ordinal β_n.

It remains to show that the strategy described above is a good winning strategy for II in $\mathscr{G}(P_\gamma)$. Let $\gamma_\infty = \lim_n \gamma_n$ and $S = \{\beta_n\}_n$. As in the proof of 5.8 we obtain a sequence $q = \langle \dot{q}_n \rangle_n$ in the ω-iteration R of $\{\dot{R}_n\}_n$ such that $q \leqslant \langle \dot{p}_n \rangle_n$ and

$$q \Vdash \forall n \exists k \quad \dot{\alpha}_n^n = \beta_k$$

Since the $\dot{\sigma}_n$ are good winning strategies, it follows, as in the proof of the ω-Factor Lemma, that $B(R) = B(P_{\gamma_\infty})$ and that q is a condition in P_{γ_∞} and support$(q) \subseteq S$. Let us identify the condition $q \in P_{\gamma_\infty}$ with $\widehat{q}\,\widehat{1}\,\widehat{1} \ldots \in P_\gamma$ (if $\gamma_\infty < \gamma$). Since $S \subseteq \gamma_\infty$ and for every n, $q \restriction n \Vdash \text{support}(\dot{s}_n) \subseteq S$, we have

$$q \leqslant (\langle p_0 \cdots \dot{p}_n \rangle, \dot{s}_n)$$

It follows that $q \leqslant p$, $q \Vdash \forall n \exists k \ \dot{\alpha}_n = \beta_k$ (and support$(q) \subseteq S$). Hence, the strategy is a good winning strategy. □

5.25 Iteration of length ω_2

By Theorem 5.1, iteration of proper forcing preserves $\aleph_1$. Assuming the CH, iteration of length at most ω_2 preserves other cardinals as well, provided the forcing notions iterated are of size at most $\aleph_1$. This has applications when one adds iteratively generic reals, such as $\aleph_2$ Laver reals. We conclude this chapter by stating (without proof) the following lemma, the consequence of which is (see Corollary 7.10, Part II) that an iteration of length ω_2 of proper forcings of size $\aleph_1$ has the $\aleph_2$-chain condition and therefore preserves all cardinals:

5.26 Lemma

Assume the CH. Let $\alpha < \omega_2$ and let P_α be a countable support iteration of proper forcings $\{\dot{Q}_\xi\}_\xi$ such that for all $\xi < \alpha$, $\Vdash_\xi |\dot{Q}_\xi| \leqslant \aleph_1$. Then P_α has a dense subset of size $\leqslant \aleph_1$, and $\Vdash_{P_\alpha}$ CH. □

Reference

Theorem 5.1 is due to Shelah. The proof presented here uses the method of C. Gray.

6 The Proper Forcing Axiom

We consider a powerful internal forcing axiom that generalizes the version $\mathrm{MA}\aleph_1$ of Martin's Axiom (Part II, Section 3.2).

6.1 Proper Forcing Axiom (PFA)

If P is a proper notion of forcing and if $\{D_\alpha:\alpha<\omega_1\}$ are dense subsets of P, then there exists a filter G on P that meets all the D_α.

The aim of this chapter is to prove the following consistency result, due to Baumgartner and Shelah:

6.2 Theorem

If there exists a supercompact cardinal then there is a generic model that satisfies PFA and $2^{\aleph_0}=\aleph_2$.

6.3

A few remarks before we start the proof of Theorem 6.2. Every c.c.c. notion of forcing is proper, and so PFA implies $\mathrm{MA}\aleph_1$. Consequently, PFA implies $2^{\aleph_0}>\aleph_1$.

At present it is unknown whether PFA is consistent with $2^{\aleph_0}>\aleph_2$, but a stronger axiom than PFA, the so-called Martin's Maximum, implies that $2^{\aleph_0}=\aleph_2$.

Unlike (the general version of) Martin's Axiom, PFA only deals with $\aleph_1$ dense sets at a time. It is impossible to generalize PFA to deal with $\aleph_2$ dense sets, as is easily seen when one considers the ω-closed notion of forcing P that adjoins a collapsing map of ω_2 onto ω_1 (with countable conditions): for each $\alpha<\omega_2$, let $D_\alpha=\{p\in P:\alpha\in\text{range }(p)\}$; no filter on P can meet all the dense sets $D_\alpha, \alpha<\omega_2$.

The consistency proof given below uses a supercompact cardinal. There are consequences of PFA showing that some large cardinal assumption is necessary for the consistency of PFA. At present it is not clear exactly how strong large cardinal assumption is needed.

We recall the definition of a supercompact cardinal:

6.4 Definition

An uncountable cardinal κ is *supercompact* if for every $\lambda\geqslant\kappa$ there exists an elementary embedding

$$j: V \to M$$

with critical point κ (i.e. κ is the least ordinal such that $j(\kappa) > \kappa$) such that $j(\kappa) > \lambda$ and $M^\lambda \subset M$, i.e. every λ-sequence $\{a_\alpha : \alpha < \lambda\} \subset M$ is in M.

A major tool in the proof of Theorem 6.2 is a *Laver function*:

6.5 Lemma

Let κ be supercompact. There exists a function $f: \kappa \to V_\kappa$ such that for every set x, for every sufficiently large λ there exists an elementary embedding $j: V \to M$ with critical point κ such that $j(\kappa) > \lambda$, $M^\lambda \subset M$ and

$$(j(f))(\kappa) = x \tag{6.6}$$

'Sufficiently large λ' means $\lambda \geqslant \kappa$ and $\lambda \geqslant |\mathrm{TC}(x)|$, where $\mathrm{TC}(x)$ is the transitive closure of x. The proof of Lemma 6.5 can be found in Laver [1978].

The proof of Theorem 6.2 follows loosely the proof of the consistency of Martin's Axiom (see Theorem 3.5, Part II). We construct a notion of forcing by countable support iteration of length κ. Each notion of forcing used in the iteration is proper, and so $\aleph_1$ is preserved. The iteration uses only forcings of size $< \kappa$, and so κ and cardinals above κ are preserved. Cardinals between $\aleph_1$ and κ are collapsed, and κ becomes $\aleph_2$; and the model satisfies $2^{\aleph_0} = \aleph_2$. The crucial property of the model is of course that it satisfies PFA. This cannot be arranged as easily as in the case of MA, because we do not have an analog of Lemma 3.4, Part II (equivalence of MA with MA*). Instead, we use a Laver function, which makes it possible to handle all potential proper forcings in only κ steps.

6.7 Proof of Theorem 6.2

Let κ be supercompact, and let f be a Laver function. We construct a countable support iteration P_κ of $\{\dot{Q}_\alpha : \alpha < \kappa\}$ as follows:

At stage α, consider the set $f(\alpha)$. In V^{P_α} we define $\dot{Q}_\alpha$ as follows: if $f(\alpha)$ is a pair $(\dot{P}, \dot{\mathscr{D}})$ of P_α-names such that $\dot{P}$ is a proper notion of forcing and $\dot{\mathscr{D}}$ is a γ-sequence of dense subsets of $\dot{P}$, for some $\gamma < \kappa$, then let $\dot{Q}_\alpha = \dot{P}$; otherwise, let $\dot{Q}_\alpha$ be the trivial forcing.

Let P_κ be the countable support iteration of the $\dot{Q}_\alpha$, and let G be generic on P_κ. We shall show that $V[G]$ satisfies PFA and $2^{\aleph_0} = \aleph_2$.

As each $\dot{Q}_\alpha$ is proper, P_κ is proper. Hence, P_κ preserves $\aleph_1$. As $f(\alpha) \in V_\kappa$ for each $\alpha < \kappa$, each P_α has size less than κ, and so (by Corollary 7.10, Part II) P_κ has the κ-chain condition. Hence, cardinals $\geqslant \kappa$ are preserved.

Also, $|P_\kappa| = \kappa$, and so $V[G]$ satisfies $2^{\aleph_0} \leqslant \kappa$. Since PFA implies that

$2^{\aleph_0} \geqslant \aleph_2$, we shall be done as soon as we prove that $V[G]$ satisfies PFA and $\kappa = \aleph_2$.

We shall prove that $V[G]$ satisfies

> If P is proper and $\mathscr{D} = \{D_\alpha : \alpha < \gamma\}, \gamma < \kappa$, is a sequence of dense sets then there is a filter F on P that meets each D_α (6.8)

The condition (6.8) clearly implies PFA in $V[G]$; let me show that it also implies $\kappa = \aleph_2$: if $\gamma < \kappa$, let P be the notion of forcing that collapses γ onto ω_1 with countable conditions. P is proper. For each $\alpha < \gamma$ let $D_\alpha = \{p \in P : \alpha \in \text{range } (p)\}$. Applying (6.8), we obtain collapsing map of γ onto ω_1. Thus, $\kappa = \aleph_2$.

In order to prove (6.8) in $V[G]$, let P be a proper forcing in $V[G]$ and let $\mathscr{D} = \{D_\alpha : \alpha < \gamma\}$ be a collection of $\gamma < \kappa$ dense subsets of P. Let $\dot{P}$ and $\dot{\mathscr{D}}$ be P_κ-names for P and $\mathscr{D}$.

Let λ be a sufficiently large cardinal (say $\lambda > 2^{2^{|P|}}$); we may also assume that $P \subset \lambda$. Since f is a Laver function, there exists an elementary embedding $j : V \to M$ with critical point κ such that $j(\kappa) > \lambda$, $M^\lambda \subset M$, and

$$(jf)(\kappa) = (\dot{P}, \dot{\mathscr{D}}) \tag{6.9}$$

6.10 Lemma

P is a proper notion of forcing in $M[G]$.

Proof P is proper in $V[G]$; using Definition 3.7 (and the remark following it), this is witnessed by some club set $C \subset [V_\eta]^\omega$ of countable models, where $\eta > 2^{|P|}$ and $\eta < \lambda$. Since $M^\lambda \subset M$ and because G is generic on P_κ that has the κ-chain condition, every λ-sequence of ordinals in $V[G]$ belongs to $M[G]$. It follows that the club C is in $M[G]$. Hence, C witnesses that P is proper in the model $M[G]$. □

Now consider the notion of forcing $j(P_\kappa)$ in M. It is an iteration (with countable support) of length $j(\kappa)$, using the Laver function $j(f)$. Since $j \restriction V_\kappa$ is the identity, the first κ stages are the same as in P_κ, i.e. $P_\kappa = (j(P_\kappa)) \restriction \kappa$. At stage κ, we have $j(f)(\kappa) = (\dot{P}, \dot{\mathscr{D}})$ and P is proper in $M[G]$, and therefore $\dot{P}$ is the forcing used at stage κ of $j(P_\kappa)$. Thus, for some $\dot{R}$,

$$j(P_\kappa) = P_\kappa * \dot{P} * \dot{R} \tag{6.11}$$

Let $H * K$ be a $V[G]$-generic ultrafilter on $\dot{P} * \dot{R}$, and let us work in $V[G * H * K]$. We extend the elementary embedding $j : V \to M$ to an elementary embedding

$$j^* : V[G] \to M[G * H * K] \tag{6.12}$$

This is done as follows: for every P_κ-name $\dot{x}$, let

$$j^*(\dot{x}/G) = j(\dot{x})/(G * H * K) \tag{6.13}$$

The definition of j^* does not depend on the choice of the name $\dot{x}$, for $\|\dot{x} = \dot{y}\| \in G$ implies $\|j(\dot{x}) = j(\dot{y})\| \in G * H * K$. (Because $j(p) = p$ for every $p \in P_\kappa$.) Similarly, j^* is elementary, since $\|\varphi(\dot{x})\| \in G$ implies $\|\varphi(j(\dot{x}))\| \in G * H * K$. Also, $j^* \supset j$.

The filter H on P is $V[G]$-generic and thus it meets every $D_\alpha, \alpha < \gamma$. Let

$$F = \{j^*(p) : p \in H\} \tag{6.14}$$

As we assume that $P \subset \lambda$, we have $F = \{j(p) : p \in H\}$, and because $j \restriction \lambda \in M$, the set F is in $M[G * H * K]$. Moreover, F generates a filter on $j^*(P)$ that meets every $j^*(D_\alpha)$, $\alpha < \gamma$. Thus,

$$M[G * H * K] \Vdash \exists F \text{ on } j^*(P) \text{ that meets every } D \in j^*(\mathscr{D})$$

and, since j^* is elementary,

$$V[G] \Vdash \exists F \text{ on } P \text{ that meets every } D \in \mathscr{D}$$

which proves (6.8). □

References

Baumgartner [1983] gives a number of applications of the Proper Forcing Axiom.

Laver, R. (1978). *Israel J. Math.* **29**, 385–8.

Baumgartner, J. (1983). In *Surveys in Set Theory* (Mathias, A.R.D., ed.), Cambridge University Press, 1–59.

Shelah, S. (1982). *Proper Forcing.* Lecture Notes in Mathematics 940, Springer-Verlag, NY.

7 Martin's Maximum

The present chapter deals with a generalization of the Proper Forcing Axiom that, in a sense explained below, is a maximal consistent generalization of Martin's Axiom $MA\aleph_1$.

The axiom PFA of 6.1 is an internal forcing axiom that generalizes $MA_{\aleph_1}$. The generalization is obtained by considering a more extensive class of forcings to which the axiom applies. To explore further generalizations, let $\mathscr{C}$ be an arbitrary class of forcing notions, and let us consider the principle

7.1 MA($\mathscr{C}$):

If P is a notion of forcing in $\mathscr{C}$ and if $\{D_\alpha : \alpha < \omega_1\}$ are dense subsets of P, then there exists a filter G on P that meets all the D_α:

Thus $MA_{\aleph_1}$ is MA(c.c.c.) and PFA is MA (proper). An important generalization of PFA is *Martin's Maximum.* We say that P is *stationary preserving* if every stationary subset S of ω_1 remains stationary in $V[G]$, for any generic filter on P. Martin's Maximum is the principle MA (stationary preserving):

7.2 Martin's Maximum (MM)

If P is a stationary preserving notion of forcing and if $\{D_\alpha : \alpha < \omega_1\}$ are dense subsets of P, then there exists a filter G on P that meets all the D_α.

Since every proper notion of forcing preserves stationary sets, the class of all stationary preserving P is at least as large as the class of all proper P, and so MM implies PFA.

The maximality of the principle MM follows from this observation:

7.3 Lemma

Let P be a notion of forcing with the property that there is a stationary set $S \subseteq \omega_1$ such that

$$\| S \text{ is not stationary} \| = 1$$

Then there are $\aleph_1$ dense sets such that no filter G on P meets them all.

Proof Let $\dot{C}$ be a name such that

$$\| \dot{C} \text{ is a club and } \dot{C} \cap S = \varnothing \| = 1 \tag{7.4}$$

We define dense sets D_α and E_α, $\alpha < \omega_1$, such that no filter G can meet all the D_α and all the E_α. For each α, let

$$D_\alpha = \{p : \exists \beta \geq \alpha \quad p \Vdash \beta \in \dot{C}\}$$

$$E_\alpha = \{p\text{: either } p \Vdash \alpha \in \dot{C}, \text{ or } \exists \gamma < \alpha \text{ such that for all } \xi \text{ between } \gamma \text{ and } \alpha, \quad p \Vdash \xi \notin \dot{C}\}$$

The sets D_α and E_α are dense, because $\dot{C}$ is, respectively, unbounded and closed. Let G be a filter on P that meets all the D_α and E_α and let

$$C = \{\alpha : \exists p \in G \quad p \Vdash \alpha \in \dot{C}\}$$

The set C is unbounded because $G \cap D_\alpha \neq \emptyset$ for all α. Also, C is closed: Let α be a limit point of C and let $p \in G \cap E_\alpha$. It must be the case that $p \Vdash \alpha \in \dot{C}$, because otherwise there is $\gamma < \alpha$ such that $p \Vdash \xi \notin \dot{C}$ for all ξ between γ and α, while some other $q \in G$ forces $\xi \in \dot{C}$ for some ξ between γ and α. But p and q are compatible.

Now S is stationary, and so there is some $\alpha \in S \cap C$. Hence, some condition forces $\alpha \in \dot{C}$ for an $\alpha \in S$ which contradicts (7.4). □

The following theorem of Foreman, Magidor and Shelah establishes the consistency of Martin's Maximum (relative to a supercompact cardinal):

7.5 Theorem

If there exists a supercompact cardinal then there is a generic model that satisfies MM.

I will not give a proof of the consistency of MM here. Instead, an outline of the proof, as well as some discussion of the technique, is given below.

In the second half of this chapter, we give three applications of MM. One consequence of MM is that $2^{\aleph_0} = \aleph_2$. This underscores the still open problem whether PFA is consistent with $2^{\aleph_0} > \aleph_2$. Another consequence of MM is that the club filter on $\aleph_1$ is $\aleph_2$-saturated. This is known to imply strong large cardinal properties, thus justifying the use of a supercompact cardinal in the consistency proof of MM (and possibly of PFA as well).

7.6

An obvious attempt at a generalization of the consistency proof of PFA given in Chapter 6 would be to iterate notions of forcing that are stationary preserving. Such an approach does not succeed, for the following reason. There is a stationary preserving notion of forcing P_f, that for a given sufficiently fast growing function $f : \omega_1 \to \omega_1$, adjoints generically a fast growing function g which is eventually below f. It is clear that such forcing cannot be iterated ω times without collapsing $\aleph_1$. (For this example, consult [Shelah, 1982], p. 255.)

Semiproper forcing

We consider a property of forcing intermediate between proper and stationary preserving. For motivation, we refer the reader to Chapter 3, particularly to Theorem 3.13 showing that Definitions 3.1, 3.2, 3.7 and 3.11 are equivalent.

7.7 A game theoretic definition

Consider the *semiproper game* on P. Player I selects a condition p, and chooses a name $\dot{\alpha}_0$ for a countable ordinal. Player II chooses an ordinal β_0. At the nth move, I plays a name $\dot{\alpha}_n$ for a countable ordinal, and II plays an ordinal β_n:

$$p; \dot{\alpha}_0, \beta_0, \dot{\alpha}_1, \beta_1, \ldots, \dot{\alpha}_n, \beta_n, \ldots \tag{7.8}$$

II wins the game if and only if

$$\exists q \leqslant p \quad \forall n \quad q \Vdash \exists k \quad \dot{\alpha}_n = \beta_k \tag{7.9}$$

7.10 Definition

A notion of forcing $(P, <)$ is *semiproper* if player II has a winning strategy in the semiproper game for P.

7.11 A model theoretic definition

Let λ be sufficiently large and let M be an elementary submodel of $(V_\lambda, \in, P, <)$. A condition q is (P, M)-*semigeneric* if for every name $\dot{\alpha} \in M$ for a countable ordinal,

$$q \Vdash \exists \beta \in M \quad \dot{\alpha} = \beta \tag{7.12}$$

7.13 Definition

A notion of forcing $(P, <)$ is *semiproper* if for some (for all) sufficiently large λ there is a club set $C \subseteq [V_\lambda]^\omega$ of countable elementary submodels $M \prec (V_\lambda, \in, P, <)$ with the following property:

$$\forall p \in M \exists q \leqslant p \quad q \text{ is } (P, M)\text{-semigeneric} \tag{7.14}$$

Definitions 7.10 and 7.13 are weaker versions of the corresponding definitions of properness. Instead of arbitrary ordinal names, the definitions refer only to names for countable ordinals.

A careful perusal of the equivalence proof in Chapter 3 will lead the reader to the following conclusion:

7.15 Theorem

(*a*) The two definitions of semiproperness given above are equivalent.

(*b*) If P is semiproper then P is stationary preserving. □

The last part of the proof of Theorem 3.13 does not go through in the semiproper case, as the argument uses preservation of stationary subsets of $[\lambda]^\omega$ for some large λ, not just $\lambda = \aleph_1$.

7.16

It is consistent, relative to a large cardinal, that every stationary preserving P is semiproper (the converse of 7.15 (*b*)); we discuss this later in some detail. As shown by Shelah, a large cardinal assumption is necessary for this implication. There is a notion of forcing (the *Namba forcing*) which is stationary preserving, but not semiproper unless $0^\#$ exists.

Semiproper Forcing Axiom and revised countable support iteration

As 'semiproper' is a property intermediate between 'stationary preserving' and 'proper', the forcing axiom MA(semiproper) is stronger than PFA and a consequence of MM:

7.17 Semiproper Forcing Axiom (SPFA)

If P is semiproper and if $\{D_\alpha : \alpha < \omega_1\}$ are dense in P, then there is a filter G on P that meets all the D_α.

For the record*:

$$\text{MM} \rightarrow \text{SPFA} \rightarrow \text{PFA} \tag{7.18}$$

The axiom SPFA is consistent relative to a supercompact cardinal. Its consistency (due to Shelah) is established in a manner similar to the proof of Theorem 6.2, namely by iteration of semiproper forcings.

The iteration used in the consistency proof of SPFA is called *revised countable support iteration.* If one iterates proper forcing (with countable support), every countable set of ordinals at any stage α of the iteration is included in a countable set in the ground model. This fact is essential in the proof of the Factor Lemma, which enables us to replace a condition with countable support in $V[G_\alpha]$ by a condition with countable support in V.

* S. Shelah proved recently that SPFA and MM are equivalent.

When forcing with a semiproper partial order, however, the cofinality of an ordinal may change to ω. With this in mind, the method of revised countable support uses a different kind of limit at the limit stages of the iteration, in a way anticipating when an ordinal changes its cofinality to ω. Roughly speaking, the support of a condition is not a countable set of ordinals, but rather a countable set of potential names for ordinals. The execution of this idea is technically quite involved, and can be found in [Shelah, 1982].

An analog of Theorem 5.1 states that semiproperness is preserved under revised countable support iteration. This fact, along with the technique used in the proof of Theorem 6.2, establishes the consistency of SPFA relative to a supercompact cardinal.

7.19 Forcing principles $\mathrm{MA}^+(\mathscr{C})$

We shall now consider a stronger version of the internal forcing axioms (7.1):

$\mathrm{MA}^+(\mathscr{C})$:

If P is a notion of forcing in $\mathscr{C}$, if $\{D_\alpha : \alpha < \omega_1\}$ are dense subsets of P and if $\dot{S}$ is a name such that

$$\| \dot{S} \text{ is a stationary subset of } \omega_1 \| = 1$$

then there exists a filter G on P that meets all the D_α, and such that the set

$$\dot{S}/G = \{\alpha : \exists p \in G \quad p \Vdash \alpha \in \dot{S}\}$$

is stationary. (7.20)

For the particular classes $\mathscr{C}$ we use the notation $\mathrm{MA}^+_{\aleph_1}$, PFA^+, SPFA^+ and MM^+. Baumgartner has shown that $\mathrm{MA}^+_{\aleph_1} = \mathrm{MA}_{\aleph_1}$; the plussed version of the other forcing axioms is strictly stronger than the axioms themselves. The consistency proof of PFA in Chapter 6 can be modified in the obvious way to yield the consistency of PFA^+. Similarly, the consistency proof of SPFA can be modified to give the consistency of SPFA^+. Thus:

7.21 Theorem

SPFA^+ is consistent relative to a supercompact cardinal.

We shall conclude our discussion on the consistency of MM by proving the following:

7.22 Theorem

$\mathrm{MA}^+(\omega\text{-closed})$ implies that every stationary preserving forcing P is semiproper.

7.23 Corollary

SPFA$^+ \leftrightarrow$ MM$^+$

And consequently, Theorem 7.21 yields Theorem 7.5, the consistency of Martin's Maximum.

Note that by 7.16, the axiom MA$^+$(ω-closed) implies (at least) $0^\#$.

Toward the proof of Theorem 7.22, let P be a notion of forcing, let λ be a sufficiently large cardinal, and consider the club set Γ of all countable elementary submodels of the model $(V_\lambda, \in, P, <)$. An *elementary chain* is a sequence

$$\langle M_\alpha : \alpha < \theta \rangle \quad (\theta \leqslant \omega_1) \tag{7.24}$$

of elements of Γ such that $M_\alpha \prec M_\beta$ whenever $\alpha < \beta$, and $M_\gamma = \bigcup_{\alpha<\gamma} M_\alpha$ for every limit γ.

7.25 Lemma

MA$^+$(ω-closed) implies that for every stationary set $X \subseteq \Gamma$ there exists an elementary chain $\langle M_\alpha : \alpha < \omega_1 \rangle$ such that $M_\alpha \supseteq \alpha$ for all α, and that the set

$$\{\alpha : M_\alpha \in X\}$$

is stationary.

Proof Let Q be the following ω-closed notion of forcing: a condition q is an elementary chain $\langle M_\alpha : \alpha < \theta \rangle$ for some countable θ, such that $M_\alpha \supseteq \alpha$ for all α; the ordering is by extension. A generic filter G on Q produces an elementary ω_1-chain $C_G = \langle M_G(\alpha) : \alpha < \omega_1 \rangle$ whose union is all of V_λ (in fact, Q is isomorphic to the usual collapse of V_λ onto $\aleph_1$). The set C_G is a club subset of Γ, and since X remains a stationary subset of Γ, the set $S_G = \{\alpha : M_G(\alpha) \in X\}$ is a stationary subset of ω_1 in $V[G]$. Let $\mathring{S}$ be the canonical name for S_G.

By the assumption, there is a filter G on Q that meets every $D_\alpha = \{q \in Q : \text{length}(q) \geqslant \alpha\}$, and such that $\mathring{S}/G$ is stationary. The filter G produces an elementary chain $\langle M_\alpha : \alpha < \omega_1 \rangle$ with the required properties. □

Proof of Theorem 7.22

Assume MA$^+$(ω-closed). Let P be a stationary preserving notion of forcing, and assume that P is not semiproper. Let λ be sufficiently large; we assume that the set of all $M \prec (V_\lambda, \in, P, <)$ without property (7.14) is stationary. By normality of the club filter, there is a condition p such that the set

$$X_p = \{M \in \Gamma : p \in M \quad \text{and} \quad \forall q \leq p \quad q \text{ is not } (P, M)\text{-semigeneric}\}$$

is stationary. (7.26)

By Lemma 7.25 there exists an elementary chain $\langle M_\alpha : \alpha < \omega_1 \rangle$ in Γ such that $M_\alpha \supseteq \alpha$ and that the set $S = \{\alpha : M_\alpha \in X_p\}$ is stationary.

We shall show that if G is a generic filter on P with $p \in G$, then, in $V[G]$, the set $\{\alpha : M_\alpha \in X_p\}$ is nonstationary. This is a contradiction, as P is stationary preserving.

For each $M \in \Gamma$, let

$$b_M = p \cdot \sum_{B(P)} \{r \in P : r \text{ is } (P, M)\text{-semigeneric}\}$$

(b_M is an element of $B(P)$, and is possibly 0). Let G be a generic filter on P with $p \in G$; we shall show that the set $\{\alpha : b_{M_\alpha} \in G\}$ contains a club.

Let $\mathring{\delta}_\xi$, $\xi < \omega_1$, enumerate (in V) all the names $\mathring{\delta}$ for countable ordinals such that $\mathring{\delta} \in \bigcup_{\alpha < \omega_1} M_\alpha$. The set

$$C_0 = \{\alpha : M_\alpha \cap \omega_1 = \alpha \quad \text{and} \quad (\forall \xi < \alpha) \mathring{\delta}_\xi \in M_\alpha\}$$

is a club (in V). Let, in $V[G]$,

$$C = \{\alpha \in C_0 : (\forall \xi < \alpha) \mathring{\delta}_\xi / G < \alpha\}$$

C is a club, and if $\alpha \in C$ then for every $\mathring{\delta} \in M_\alpha$ there is $q \in G$ such that $q \Vdash (\exists \beta \in M_\alpha) \mathring{\delta} = \beta$. Hence, $b_{M_\alpha} \in G$. Consequently, $C \subseteq \{\alpha : b_{M_\alpha} \in G\}$. □

Consequences of Martin's Maximum

Of the numerous applications of Martin's Maximum, we present three here: the continuum is equal to $\aleph_2$, the Singular Cardinals Hypothesis holds, and the club filter on $\aleph_1$ is $\aleph_2$-saturated.

7.27 Theorem

Assume MM. If $\kappa \geq \aleph_2$ is a regular cardinal then $\kappa^{\aleph_0} = \kappa$.

7.28 Corollary

MM implies $2^{\aleph_0} = \aleph_2$.

7.29 Corollary

MM implies the Singular Cardinals Hypothesis: for every singular λ, if $2^{\operatorname{cf} \lambda} < \lambda$ then $\lambda^{\operatorname{cf} \lambda} = \lambda^+$.

(By Silver's theorem, it suffices to prove Corollary 7.29 for the case $\mathrm{cf}\,\lambda = \omega$; this follows from Theorem 7.27 by letting $\kappa = \lambda^+$.)

Theorem 7.27 uses the following lemma:

7.30 Lemma

Assume MM, and let $\kappa \geqslant \aleph_2$ be regular. Let $E = \{\xi < \kappa: \mathrm{cf}\,\xi = \omega\}$, and let S_n, $n \in \omega$, be disjoint stationary subsets of E. Then there exists an ordinal θ of cofinality ω_1 such that $\theta \cap \bigcup_{n=0}^{\infty} S_n$ contains a club subset of θ, and every $\theta \cap S_n$ is a stationary subset of θ.

First we prove that Lemma 7.30 implies the theorem: let $\{S_i: i < \kappa\}$ be a pairwise disjoint collection of stationary subsets of κ. Given a countable set $X \subset \kappa$, Lemma 7.30 provides an ordinal $\theta_X < \kappa$ such that if $i \in X$ then $S_i \cap \theta_X$ is stationary in θ_X, and if $i \notin X$ then $S_i \cap \theta_X$ is not stationary in θ_X. Thus, the function $X \mapsto \theta_X$ is one-to-one, and we have $\kappa^{\aleph_0} = \kappa$.

Proof of Lemma 7.30

Let S_n, $n < \omega$, be given, and let $S = \bigcup_{n=0}^{\infty} S_n$. To find the ordinal θ, we apply MM to a certain stationary preserving notion of forcing P.

Let us fix a partition of ω_1 into disjoint stationary sets $\{A_n: n < \omega\}$. A forcing condition in P is a continuous increasing function $\langle f(\nu): \nu \leqslant \alpha \rangle$ of countable length $\alpha + 1$ with values in S, such that for every $\nu \leqslant \alpha$

$$\text{if} \quad \nu \in A_n \quad \text{then} \quad f(\nu) \in S_n \tag{7.31}$$

The ordering of P is by extension.

Once we prove that P is stationary preserving, and that for every $\alpha < \omega_1$ the set $D_\alpha = \{f \in P: \mathrm{dom}\, f \geqslant \alpha + 1\}$ is dense in P, Martin's Maximum yields a continuous increasing function $f: \omega_1 \to S$ with property (7.31), and $\theta = \sup_\nu f(\nu)$ is as desired.

In both proofs we shall make use of the following:

7.32 Lemma

Let $A \subseteq \omega_1$ be stationary, and let T be a stationary subset of κ such that $\mathrm{cf}\,\xi = \omega$ for all $\xi \in T$. Every model $(V_\lambda, \in, \ldots)$, where $\lambda \geqslant \kappa$, has a countable elementary submodel M such that $M \cap \omega_1 \in A$ and $\sup(M \cap \kappa) \in T$.

Proof Since T is stationary in κ, there exists an elementary submodel N (of size $\aleph_1$) such that $\omega_1 \subset N$ and $\sup(N \cap \kappa) \in T$. N is the union of an elementary ω_1-chain of countable models M_α such that $\sup(M_\alpha \cap \kappa) = \sup(N \cap \kappa)$ for each α. Since A is stationary in $\aleph_1$, there is $M = M_\alpha$ such that $M \cap \omega_1 \in S$. □

We prove first that each D_α is dense. We do it by induction on α. Assume that it is true for all ordinals less than α. Let $p \in P$; we find $q \supseteq p$ of length at least $\alpha + 1$. Let n be such that $\alpha \in A_n$, and let $T = S_n$. By Lemma 7.32 there is a countable elementary submodel M of $(V_\lambda, P, p, \alpha)$ for sufficiently large λ such that $\gamma = \sup(M \cap \kappa) \in T$.

Let $\{\alpha_n\}_n$ be a sequence of countable ordinals with limit α, and let $\{\gamma_n\}_n$ be a sequence converging to γ. We construct a sequence of conditions $p = p_0 \subset p_1 \subset \cdots \subset p_n \subset \cdots$ as follows, such that $p_n \in M$ for each n: given p_n, let p_{n+1} be, inside M, an extension of p_n such that $\alpha_n \in \mathrm{dom}(p_{n+1})$ and that $p_{n+1}(\alpha_n) \geqslant \gamma_n$; such p_{n+1} exists because M, an elementary submodel of V_λ, satisfies the induction hypothesis that D_{α_n} is dense. The function $q = \bigcup_{n=0}^{\infty} p_n \cup \{(\alpha, \gamma)\}$ is a condition of length $\alpha + 1$.

Finally, we prove that P is stationary preserving. Let A be a stationary subset of ω_1, let $p \in P$, and let $\dot{C}$ be a name such that

$$p \Vdash \dot{C} \text{ is a club subset of } \omega_1 \tag{7.33}$$

We want a stronger q and some $\alpha \in A$ such that $q \Vdash \alpha \in \dot{C}$.

Let n be such that $A \cap A_n$ is stationary, and let $T = S_n$. By Lemma 7.32 there is a countable elementary submodel M of $(V_\lambda, P, p, \dot{C})$ such that $\alpha = M \cap \omega_1 \in A \cap A_n$ and $\gamma = \sup(M \cap \kappa) \in T$.

Let $\{\alpha_n\}_n$ be a sequence with limit α, and let $\{\gamma_n\}_n$ be a sequence with limit γ. We construct conditions $p = p_0 \subset \cdots \subset p_n \subset \cdots$, each in M, as follows: given p_n, let $p_{n+1} \in M$ be such that $\alpha_n \in \mathrm{dom}(p_{n+1})$, that $p_{n+1}(\alpha_n) \geqslant \gamma_n$, and that for some $\beta_n \geqslant \alpha_n$ in M (therefore $< \alpha$), $p_{n+1} \Vdash \beta_n \in C$.

The function $q = \bigcup_{n=0}^{\infty} p_n \cup \{(\alpha, \gamma)\}$ is a condition, and it follows that $q \Vdash \alpha \in \dot{C}$.

This concludes the proof of Lemma 7.30, and of Theorem 7.27 as well. □

7.34 Definition

The club filter on $\aleph_1$ is $\aleph_2$-*saturated* if there is no collection of stationary sets $\{A_i : i < \aleph_2\}$ such that $A_i \cap A_j$ is nonstationary for all $i \neq j$.

By the work of Solovay, Kunen and Mitchell, $\aleph_2$-saturation of the club filter has very strong large cardinal consequences. By the following theorem, so does Martin's Maximum:

7.35 Theorem

MM implies that the club filter on $\aleph_1$ is $\aleph_2$-saturated.

We start the proof by reviewing some facts on stationary sets of countable ordinals.

7.36 Lemma

For any collection $\{A_i : i \in Z\}$ of stationary sets, with $|Z| = \aleph_1$, there is a stationary set A such that

(i) $A_i - A$ is nonstationary, for all $i \in Z$, and
(ii) every stationary subset of Z has stationary intersection with some A_i.

The set A is unique modulo the club filter, and is denoted $A = \sum_{i \in Z} A_i$.

Proof Assuming $Z = \omega_1$, we let A be the diagonal union:

$$\alpha \in A \leftrightarrow \alpha \in \bigcup_{\xi < \alpha} A_\xi \quad \square \tag{7.37}$$

Let S be a stationary set and let P_S be the notion of forcing that adds a club subset of S (Example 1.2).

7.38 Lemma

If $A \subseteq S$ is stationary, then A remains stationary in V^{P_S}.

Proof We follow closely the proof of ω-distributivity (Lemma 1.4). Let

$$p \Vdash \dot{C} \text{ is a club}$$

we shall find $q \leqslant p$ and $\lambda \in A$ such that $q \Vdash \lambda \in \dot{C}$.

As in Lemma 1.4, we construct a chain $\{A_\alpha\}_\alpha$ of countable subsets of P_S and an increasing sequence γ_α. Given A_α, for each $q \in A_\alpha$ we find $\beta = \beta(q) > \gamma_\alpha$ and some $r = r(q) \leqslant q$ so that $r \Vdash \beta \in \dot{C}$, and $\max(r) > \gamma_\alpha$. Then we let $A_{\alpha+1} = A_\alpha \cup \{r(q) : q \in A_\alpha\}$ and $\gamma_{\alpha+1} = \sup\{\max(r) : r \in A_{\alpha+1}\}$.

As in Lemma 1.4, we find $\lambda \in A$ with the property that $\gamma_\alpha < \lambda$ whenever $\alpha < \lambda$. Then we find a sequence of conditions $p_0 \geqslant p_1 \geqslant \cdots \geqslant p_n \geqslant \cdots$ such that both $\max(p_n)$ and $\beta(p_n)$ converge to λ, and consequently, $q = \bigcup_n p_n \cup \{\lambda\}$ forces that $\lambda \in \dot{C}$. $\square$

We are now ready to prove Theorem 7.35:

7.39 Lemma

Assume MM, and let $\{A_i : i \in W\}$ be a collection of stationary sets with the property that every stationary set has stationary intersection with some A_i. Then there is $Z \subseteq W$ of size $\leqslant \aleph_1$ such that $\sum_{i \in Z} A_i$ is in the club filter; hence. every stationary set has stationary intersection with some A_i, $i \in Z$.

The lemma clearly implies the theorem.

Proof We apply MM to the following forcing notion P: first let Q be the forcing that collapses $|W|$ onto $\aleph_1$ with countable conditions. In V^Q, consider $\dot{S} = \sum_{i \in W} A_i$ and let $P = Q * P_{\dot{S}}$.

Equivalently, let P (a dense set in $Q * P_{\dot{S}}$) be the set of all pairs (q, p) such that

$$\begin{aligned} &q\colon\ \gamma + 1 \to W \text{ for some } \gamma < \omega_1, \text{ and} \\ &p \text{ is a closed countable subset of } \omega_1 \text{ such that} \\ &\alpha \in p \text{ implies } \alpha \in \bigcup_{\xi < \alpha} A_{q(\xi)} \end{aligned} \tag{7.40}$$

A condition (q', p') is stronger than (q, p) if $q' \supseteq q$ and p' is an end-extension of p.

Note that P is stationary preserving. If A is stationary, then for some $i \in W$, $A \cap A_i$ is stationary. $A \cap A_i$ remains stationary in V^Q, and so $A \cap A_i \cap \dot{S}$ is stationary in V^Q. By Lemma 7.38, $A \cap A_i \cap \dot{S}$ remains stationary in $V^{Q * P_{\dot{S}}}$. Hence, A is stationary in V^P.

For each $\alpha < \omega_1$, let $D_\alpha = \{(q, p) \in P : \alpha \leqslant \max(p)\}$. Each D_α is dense in P. By Martin's maximum, there is a filter G on P that meets all the D_α. Let

$$F = \bigcup \{q : (q, p) \in G \text{ for some } p\}$$
$$Z = \text{range}(F)$$

and

$$C = \bigcup \{p : (q, p) \in G \text{ for some } q\}$$

The set C is a club, and by (7.40)

$$\alpha \in C \leftrightarrow \alpha \in \bigcup_{\xi < \alpha} A_{F(\xi)}$$

In other words, $C = \sum_{i \in Z} A_i$. □

References

Foreman, M., Magidor, M. and Shelah, S. (1986). *Ann. Math.* (in press).
Shelah, S. (1982). *Proper Forcing*. Lecture Notes in Mathematics 940, Springer-Verlag, NY.

8 *Well-founded iteration*

In this final chapter, we discuss some of the ramifications of the iteration methods described in earlier chapters, as well as questions raised by the consistency proof of the Proper Forcing Axiom.

One still unresolved question is whether the proper forcing axiom is consistent with $2^{\aleph_0} > \aleph_2$. In view of Martin's Maximum, which implies $2^{\aleph_0} = \aleph_2$, this question may not be just a technical problem resulting from the particular method used in the consistency proof. Nevertheless, the inability to construct a model with $2^{\aleph_0} > \aleph_2$ by the method of Chapter 6 is caused by the failure of the countable support iteration to preserve cardinals above $\aleph_1$. Indeed, it has been observed that countable support iterations of length $> \omega_2$ do not necessarily preserve $\aleph_2$. One effect of this phenomenon is that in all applications of the countable support iteration one ends up with a model in which the continuum is exactly $\aleph_2$. This is in contrast with the finite support iteration which, as in the case of Martin's Axiom, enables us to construct models with the continuum arbitrarily large.

We shall describe a more general method of iteration that makes it possible to preserve cardinals while adding a large number of generic reals. The method is a generalization of both iterated forcing and product forcing.

In particular, it is then possible to give a partial answer to the question of consistency of PFA with $2^{\aleph_0} > \aleph_2$. Without going into details, let me state the result as follows. There is a class $\mathscr{B}$ of proper forcing notions that includes, among others, Cohen forcing, Sacks forcing, Prikry–Silver forcing, Mathias forcing and Laver forcing, and for which the analog of PFA is consistent with $2^{\aleph_0} > \aleph_2$.

8.1 Theorem

The axiom MA($\mathscr{B}$) is consistent with $2^{\aleph_0} > \aleph_2$.

The class $\mathscr{B}$ is considerably smaller than the class of all proper forcings. Note that there is no mention of large cardinals in Theorem 8.1, as the consistency is relative just to ZFC. This is unlike the consistency of PFA for which large cardinals are necessary.

We omit the proof of Theorem 8.1 but outline the method of well-founded iteration which the proof uses.

Well-founded iteration

The following definition is a generalization of iteration of forcing described in Definition 7.1, Part II.

Let $(W, <)$ be a well-founded partial order. For $x \in W$, let $W[x]$ denote the initial segment $\{y \in W: y < x\}$. In general, an *initial segment* of W is a set $X \subseteq W$ with the property that $x \in X$ and $y < x$ then $y \in X$. Definition 8.2 is by induction on the height of W:

8.2 Definition

Let $(W, <)$ be a well-founded partial order. A forcing notion P_W is a (*well-founded*) *iteration along* W if it is a set of functions on W with the following properties:

(i) For every $x \in W$, $P_{W[x]} = P_W \upharpoonright W[x] = \{p \upharpoonright W[x] : p \in P_W\}$ is an iteration along $W[x]$

(ii) For every $x \in W$ there is a forcing notion $\dot{Q}_x \in V^{P_{W[x]}}$, and for every $p \in P_W$

$$\Vdash_{P_{W[x]}} p(x) \in \dot{Q}_x$$

(iii) For all $p, q \in P_W$

$$p \leqslant_{P_W} q \quad \text{iff for all } x \in W$$

$$p \upharpoonright W[x] \leqslant_{P_{W[x]}} q \upharpoonright W[x]$$

and

$$p \upharpoonright W[x] \Vdash_{P_{W[x]}} p(x) \leqslant_{\dot{Q}_x} q(x)$$

(iv) $1 \in P_W$, where $1(x) = 1$ for all $x \in W$.
If X is an initial segment of W, let $P_X = P_W \upharpoonright X = \{p \upharpoonright X : p \in P_W\}$.

(v) If X is an initial segment of W and if $p \in P_W$ and $q \in P_X$ are such that $q \leqslant_{P_X} p \upharpoonright X$ then the following function r is a condition in P_W:

$$r(x) = \begin{cases} q(x) & \text{if} \quad x \in X \\ p(x) & \text{if} \quad x \in W - X \end{cases}$$

If the well-founded partial ordering W in Definition 8.2 is a well ordering (of length α) then Definition 8.2 is just Definition 7.1, Part II, of iteration of length α. The analog of Lemma 7.2, Part II, holds for well-founded iterations as well:

8.3 Lemma

If X is an initial segment of W and $P_X = P_W \upharpoonright X$, then $V^{P_X} \subseteq V^{P_W}$.

□

Note that well-founded iteration is also a generalization of product forcing: a product of forcing notions $\{P_i: i \in I\}$ is in fact an iteration along I, where I is given the discrete partial order.

As in the case of well-ordered iteration, we can define the forcing notion P_W by giving the names $\dot{Q}_x$, $x \in W$, and specifying the *support* of each condition. In particular, we may consider well-founded iterations with finite support, with countable support, etc.

To illustrate the use of well-founded iterations, let us consider the problem of preserving $\aleph_2$. Suppose that P_W is such that the support of every $p \in P_W$ is countable, and is included in an initial segment X of w of size $\leqslant \aleph_1$. (Note that there are arbitrarily large partial orders with this property; only the height of W is limited to $\leqslant \aleph_2$.) Assuming that each $P_{W[x]}$ has the $\aleph_2$-chain condition, a standard Δ-system argument shows that P_W has the $\aleph_2$-chain condition as well.

The preservation of $\aleph_1$ by well-founded iteration is an entirely different problem. In some cases, it suffices to use countable supports. Such is the case when iterating Sacks forcing, for example; a somewhat complicated fusion argument shows that P_W is proper (in fact, satisfies Axiom A). However, when applying well-founded iteration to other forcings, such as in the proof of Theorem 8.1, one needs a more general iteration. Note that in view of Propositions 5.16 and 5.17, Part I, and because iteration along W subsumes product forcing, neither finite support iteration nor countable support iteration is suitable when iterating Laver or Mathias forcing.

The iteration used in the proof of Theorem 8.1 uses *mixed support*. We omit the (rather technical) definition of mixed support iteration, and instead give an example, a special case of mixed support iteration that is a prototype of the method.

8.4 Example (Baumgartner) Mixed support product of Mathias forcings

Let us consider the following product P of a collection $\{P_i: i \in I\}$, where each P_i is Mathias forcing (Section 3.11, Part I). Condition in P are certain functions $p = \langle p(i): i \in I \rangle$ such that each $p(i)$ is a Mathias condition (s_i, S_i). The *support* of p is the set

$$\operatorname{supp}(\langle p_i: i \in I \rangle) = \{i: p_i \neq 1\}$$

(Here $1 = (\varnothing, \omega)$.) The *root* of p is the set

$$\operatorname{root}(\langle p_i: i \in I \rangle) = \{i: \operatorname{stem}(p_i) \neq \varnothing\}$$

8.5 Definition

The *mixed support product* of Mathias forcings $\{P_i : i \in I\}$ is the set of all functions $p = \langle p_i : i \in I \rangle$ whose support is countable and whose root is finite.

The mixed support product of Mathias forcing preserves $\aleph_1$:

8.6 Proposition

If $p \Vdash \dot{X} : \omega \to V$ then there exists $q \leqslant p$ and a countable A such that $q \Vdash \dot{X} \subseteq A$. Moreover, we can find the $q \leqslant p$ with the same root as p, and, in fact, $\text{stem}(q_i) = \text{stem}(p_i)$ for every $i \in I$.

Proof The proof is a variation on the proof of Lemma 3.12, Part I, or Theorem 5.12, Part I. To simplify matters, assume that p has empty root; we shall find q as desired, with empty root as well.

Let $\{u_n\}_n$ be a sequence of natural numbers such that each u appears infinitely often. We construct a sequence $p = p_0 \geqslant p_1 \geqslant p_2 \geqslant \cdots$; let W_n denote the support of p_n. Let $W = \bigcup_{n=0}^{\infty} W_n$; by a suitable enumeration we also construct a sequence of finite sets $F_0 \subseteq F_1 \subseteq F_2 \subseteq \cdots$ with $\bigcup_{n=0}^{\infty} F_n = W$. For every n, we make sure that

$$\forall i \in F_n \quad p_{n+1}(i) \leqslant_n p_n(i)$$

This guarantees that for each $i \in W$, $p_\infty(i) = \lim_n p_n(i)$ is a Mathias condition (with empty stem), and so $p_\infty = \langle p_\infty(i) \rangle_i$ is a condition with support W and empty root. We also produce a sequence of finite sets $\{A_n\}_n$ with the iteration that $p_\infty \Vdash \dot{X} \subseteq A$, where $A = \bigcup_{n=0}^{\infty} A_n$.

At stage n, we already have $p_n = \langle (\varnothing, S_n^\bullet(i)) : i \in I \rangle$. For each $i \in F_n$, let $K_n(i)$ be the set of first n elements of $S_n(i)$. Let $\tau_1, \ldots, \tau_l$ be all the functions τ on F_n such that $\tau = \langle t(i) : i \in I \rangle$, where each $t(i)$ is a finite increasing sequence with values in $S_n(i)$. We construct p_{n+1} in l steps, by constructing $p_n = q_0 \geqslant \cdots \geqslant q_l = p_{n+1}$, and also construct the finite set A_n.

At stage $k \leqslant l$, let $q_k - \langle (\varnothing, S^{(k)}(i) : i \in I \rangle$. If there exists a condition $r = \langle (t(i), T(i)) : i \in I \rangle$ with root F_n such that $\langle t(i) : i \in F_n \rangle = \tau_{k+1}$ and $T(i) \subseteq S^{(k)}(i) - K_n(i)$, and that

$$r \Vdash \dot{X}(u_n) = a_n^k \tag{8.7}$$

for some a_n^k then we put a_n^k into A_n, and let q_{k+1} be the condition (with empty root) $\langle (\varnothing, S^{(k+1)}(i)) : i \in I \rangle$, where

$$S^{(k+1)}(i) = \begin{cases} K_n(i) \cup T(i) & \text{if } i \in F_n \\ T(i) & \text{otherwise} \end{cases}$$

The sequence $\{p_n\}_n$ converges to a condition p_∞. The rest of the proof is a verification of the fact that $p_\infty \Vdash \dot{X} \subseteq A$. The argument follows closely the conclusion of the proof of Lemma 3.12, Part I. I omit it, as I believe that the readers who have been able to bear with me up to this point will be able to complete the proof on their own. □

Reference

Groszek, M. and Jech, T. Generalized Iteration of Forcing (1986) (in press).

Bibliography

Baumgartner, J. (1983). Iterated forcing. In *Surveys in Set Theory* (Mathias, A.R.D., ed.), pp. 1–59, Cambridge University Press.
Baumgartner, J. (1984). Applications of the proper forcing axiom, in *Handbook of Set-theoretical Topology* (Kunen, K. and Vaughan, J.E., eds.), pp. 913–59, North-Holland, Amsterdam.
Borel, E. (1919). Sur la classification des ensembles de mesure nulle, *Bull. Soc. Math. France* **47**, 97–125.
Cohen, P. (1966). *Set Theory and the Continuum Hypothesis*, Benjamin, NY.
Dales, H.G. (1979). A discontinuous homomorphism from $C(X)$, *Am. J. Math.* **101**, 647–734.
Dales, H.G. and Woodin, W.H. (1986). *An Introduction to Independence for Analysts*, London Mathematical Society Lecture Notes 115, Cambridge University Press.
Devlin, K. and Johnsbraten, H. (1974). *The Souslin Problem*. Lecture Notes in Mathematics 405, Springer-Verlag, NY.
Easton, W. (1970). Powers of regular cardinals, *Ann. Math. Logic* **1**, 139–78.
Eklof, P. (1976). Whitehead's problem is undecidable, *Am. Math. Monthly* **83**, 775–88.
Ellentuck, E. (1974). A new proof that analytic sets are Ramsey, *J. Symb. Logic* **39**, 163–5.
Esterle, J. (1978). Sur l'existence d'un homomorphisme discontinu de $C(K)$, *Proc. London Math. Soc.* **36**, 46–58.
Fleissner, W. (1984). The normal Moore space conjecture and large cardinals, in *Handbook of Set-theoretical Topology* (Kunen, K. and Vaughan, J.E. eds.), pp. 733–60, North-Holland, Amsterdam.
Foreman, M., Magidor, M. and Shelah, S. (1986). Martin's maximum, saturated ideals and non-regular ultrafilters, *Ann. Math.* (in press).
Fremlin, D.H. (1984). *Consequences of Martin's Axiom*, Cambridge Tracts in Mathematics 84, Cambridge University Press.
Gray, C.W. (1982). Iterated forcing from the strategic point of view, Thesis, Berkeley, CA.
Grigorieff, S. (1971). Combinatorics on ideals and forcing, *Ann. Math. Logic* **3**, 363–94.
Groszek, M. and Jech, T. (1986). Generalized iteration of forcing (in press).
Jech, T. (1971). Trees, *J. Symb. Logic* **36**, 1–14.
Jech, T. (1973). Some combinatorial problems concerning uncountable cardinals, *Ann. Math. Logic* **5**, 165–98.
Jech, T. (1978). *Set Theory*, Academic Press, NY.
Jech, T. (1984). More game theoretic properties of Boolean algebras, *Ann. Pure and Appl. Logic* **26**, 11–29.
Kueker, D. (1977). Countable approximations and Löwenheim–Skolem theorems, *Ann. Math. Logic* **11**, 57–103.
Kunen, K. (1980). *Set Theory*, North-Holland, Amsterdam.
Laver, R. (1976). On the consistency of Borel's conjecture, *Acta Math.* **137**, 151–69.
Laver, R. (1978). Making the supercompactness of κ indestructible under κ-directed closed forcing, *Israel J. Math.* **29**, 385–8.
Martin, D.A. and Solovay, R. (1970). Internal Cohen extensions, *Ann. Math. Logic* **2**, 143–78.
Mathias, A.R.D. (1977). Happy families, *Ann. Math. Logic* **12**, 59–111.
Menas, T. (1976). Consistency results concerning supercompactness, *Trans. Am. Math. Soc.* **223**, 61–91.

Rudin, M.E. (1969). Souslin's conjecture, *Am. Math. Monthly* **76**, 1113–19.

Rudin, M.E. (1977). Martin's axiom, in *Handbook of Mathematical Logic* (Barwise, J., ed.), pp. 491–501, North-Holland, Amsterdam.

Sacks, G. (1971). Forcing with perfect closed sets, in *Axiomatic Set Theory* (Scott, D., ed.), pp. 331–55, American Math. Society, Providence, RI.

Shelah, S. (1974). Infinite abelian groups, Whitehead problem and some constructions, *Israel J. Math.* **18**, 243–56.

Shelah, S. (1982). *Proper Forcing*. Lecture Notes in Mathematics 940, Springer-Verlag, NY.

Shoenfield, J. (1971). Unramified forcing, in *Axiomatic Set Theory* (Scott, D., ed.), pp. 357–82, American Math. Society, Providence, RI.

Shoenfield, J. (1975). Martin's axiom, *Am. Math. Monthly* **82**, 610–17.

Solovay, R. (1970). A model of set theory in which every set of reals is Lebesgue measurable, *Ann. Math.* **92**, 1–56.

Solovay, R. (1971). Real-valued measurable cardinals, in *Axiomatic Set Theory* (Scott, D., ed.), pp. 397–428, American Math. Society, Providence, RI.

Solovay, R. and Tennenbaum, S. (1971). Iterated Cohen extensions and Souslin's problem, *Ann. Math.* **94**, 201–45.

Souslin, M. (1920). Problème 3, *Fund. Math.* **1**, 223.

Woodin, W.H. (1987). Set theory and discontinuous homomorphisms of Banach algebras, *Memoirs Am. Math. Soc.* (in press).

Index of symbols

Subject index

Author index